Birds of Pakistan

The preparation of this field guide, and publication of English and Urdu editions, has been possible thanks to sponsorship from:

The World Bank/Netherlands Partnership Program (BNPP)

and the

Asia Bird Fund of BirdLife International

Maps prepared by WWF Pakistan

• Helm Field Guides •

Birds of Pakistan

Richard Grimmett
Tom Roberts
Tim Inskipp

Illustrated by Clive Byers, Daniel Cole, John Cox, Gerald Driessens, Carl D'Silva, Martin Elliott, Kim Franklin, Alan Harris, Peter Hayman, Craig Robson, Jan Wilczur, Tim Worfolk

Urdu edition by Aleem Ahmed Khan and Imran Khaliq, Ornithological Society of Pakistan, and M. Zafar-ul Islam

Maps prepared by Hassan Ali and Salman Ashraf, WWF Pakistan

CHRISTOPHER HELM · LONDON
YALE UNIVERSITY PRESS · NEW HAVEN

THE COLOUR PLATES
Clive Byers 73–78, 81, 82, 83 (part), 84–93
Daniel Cole 1–3, 15–17, 21 (part), 22, 68
John Cox 18, 19, 23
Gerald Driessens 14, 71, 72 (part)
Carl D'Silva 8–13, 20, 21 (part), 48–51, 52 (part), 53–57, 72 (part), 83 (part)
Martin Elliott 31 (part), 32–35
Kim Franklin 37 (part), 38
Alan Harris 36 (part), 37 (part), 39 (part), 40, 41, 45, 46, 58–67
Peter Hayman 24–30, 31 (part)
Craig Robson 69, 70, 79, 80
Jan Wilczur 4–7, 47, 52 (part)
Tim Worfolk 36 (part), 39 (part), 42–44

Front cover and title page illustrations by John Cox

© 2008 text by Richard Grimmett, Tom Roberts and Tim Inskipp
© 2008 maps by Tom Roberts
© 2008 illustrations by Clive Byers, Daniel Cole, John Cox, Gerald Driessens,
Carl D'Silva, Martin Elliott, Kim Franklin, Alan Harris, Peter Hayman, Craig Robson,
Jan Wilczur, Tim Worfolk

Photographs by Tom Roberts

Published in the United Kingdom in 2008 by Christopher Helm, an imprint of A & C Black
Publishers Ltd, 38 Soho Square, London W1D 3HB

978-0-7136-8800-9

A CIP catalogue record for this book is available from the British Library

Published in the United States of America by Yale University Press.
Library of Congress Control Number: 2008937282
ISBN 978-0-300-15249-4 (pbk.)

All rights reserved. No part of this publication may be
reproduced or used in any form or by any means – photographic,
electronic or mechanical, including photocopying, recording,
taping or information storage and retrieval systems – without
permission of the publishers.

This book is produced using paper that is made from wood grown in managed, sustainable forests. It is natural, renewable and recyclable. The logging and manufacturing
processes conform to the environmental reputations of the country of origin.

www.acblack.com

Edited and designed by D & N Publishing, Lambourn Woodlands,
Hungerford, Berkshire.

Printed in China

10 9 8 7 6 5 4 3 2 1

Contents

- INTRODUCTION 6 -
How to Use this Book 6
Descriptive Parts of a Bird 7
Geographical Setting 11
Main Habitats 11
Important Bird Species 14
Migration 15
- BIRDWATCHING AREAS 16 -
- NATIONAL ORGANISATIONS 25 -
- REGIONAL AND INTERNATIONAL ORGANISATIONS 26 -
- REFERENCES 28 -
- ACKNOWLEDGEMENTS 28 -
- GLOSSARY 29 -
- SELECTED BIBLIOGRAPHY 31 -
- FAMILY SUMMARIES 34 -
- COLOUR PLATES AND SPECIES ACCOUNTS 48 -
- APPENDIX: VAGRANTS AND EXTIRPATED SPECIES 234 -
- TABLES 241 -
- INDEX 248 -

INTRODUCTION

Pakistan is bounded by Iran and Afghanistan to the west, and its huge neighbours China to the north and India to the east. The country is bisected by the Indus river, which flows the entire length of the country before entering the Arabian Sea via the sprawling Indus delta in the south. Pakistan has a rich diversity of bird habitats, from the dry alpine and moist temperate forests of the western Himalayas to the deserts of Baluchistan and Sind. The Indus basin, where some of the earliest human civilisations were founded, is extensively irrigated and cultivated, providing a variety of man-made habitats. This diversity of habitats supports a wide variety of bird species, and some 669 have been recorded. More than 60 per cent of the country – land lying to the west of the Indus river and south from Peshawar to the Arabian Sea coasts – is Palaearctic in character, with a steppic dry montane habitat, and is very different from the rest of the Indian subcontinent.

Pakistan's avifauna is thus a fascinating mix of the Palaearctic and Oriental, blending species that are at the western and eastern limits of their distribution. The country has some 'specialities', which are a target for visiting birders, and is an excellent place to see species confined to the western Himalayas, such as the magnificent Western Tragopan. Pakistan lies at the crossroads for bird migration. In autumn and spring, birds pass through in vast numbers and great variety, mainly heading to and from the Indian subcontinent, but also to East Africa. Many species also stop over to spend the winter in the country.

This magnificent bird diversity is not matched by large numbers of resident and visiting birders. Birdwatching has yet to catch on in Pakistan, as it has in other countries on the subcontinent, and Pakistan is not as yet a popular destination for the travelling birder. Hopefully, this guide will help to change this. This is the first ever field guide to birds of Pakistan, and such guides elsewhere have helped to ignite a local interest in birds, and later their conservation and a concern for the wider environment. It would not have been possible to produce this guide without the support of the World Bank, which has sponsored both the English and Urdu editions.

The text and maps have drawn heavily on the two-volume *The Birds of Pakistan* (Roberts, 1991–1992), and readers are recommended to refer to this work, especially for detailed information on status, distribution, habits and habitat.

The northern regions covered by this book have had a troubled recent history, and territory is still disputed with neighbouring India. This guide covers such regions, including Baltistan and parts of Kashmir, that are accessible to birders in Pakistan, and nothing in the maps or species text is intended to confer any opinion or political significance.

HOW TO USE THIS BOOK

Taxonomy and Nomenclature

Taxonomy and nomenclature largely follow *An Annotated Checklist of the Birds of the Oriental Region* (Inskipp *et al.*, 1996). The sequence generally follows the same reference, although some species have been grouped out of this systematic order to enable useful comparisons to be made. In cases where differences in taxonomic opinion exist in the literature, the species limits are fully discussed in that work, to which readers requiring further information should initially refer. The taxonomy differs in several respects based on recent published works that are considered to provide justification for the new treatments. The Yellow-legged Gull *Larus cachinnans* has been split into two species: the Yellow-legged Gull *L. michahellis* in Europe and the Caspian Gull *L. cachinnans* further east (Klein and Buchheim, 1997; Liebers *et al.*, 2001). The Long-billed Vulture

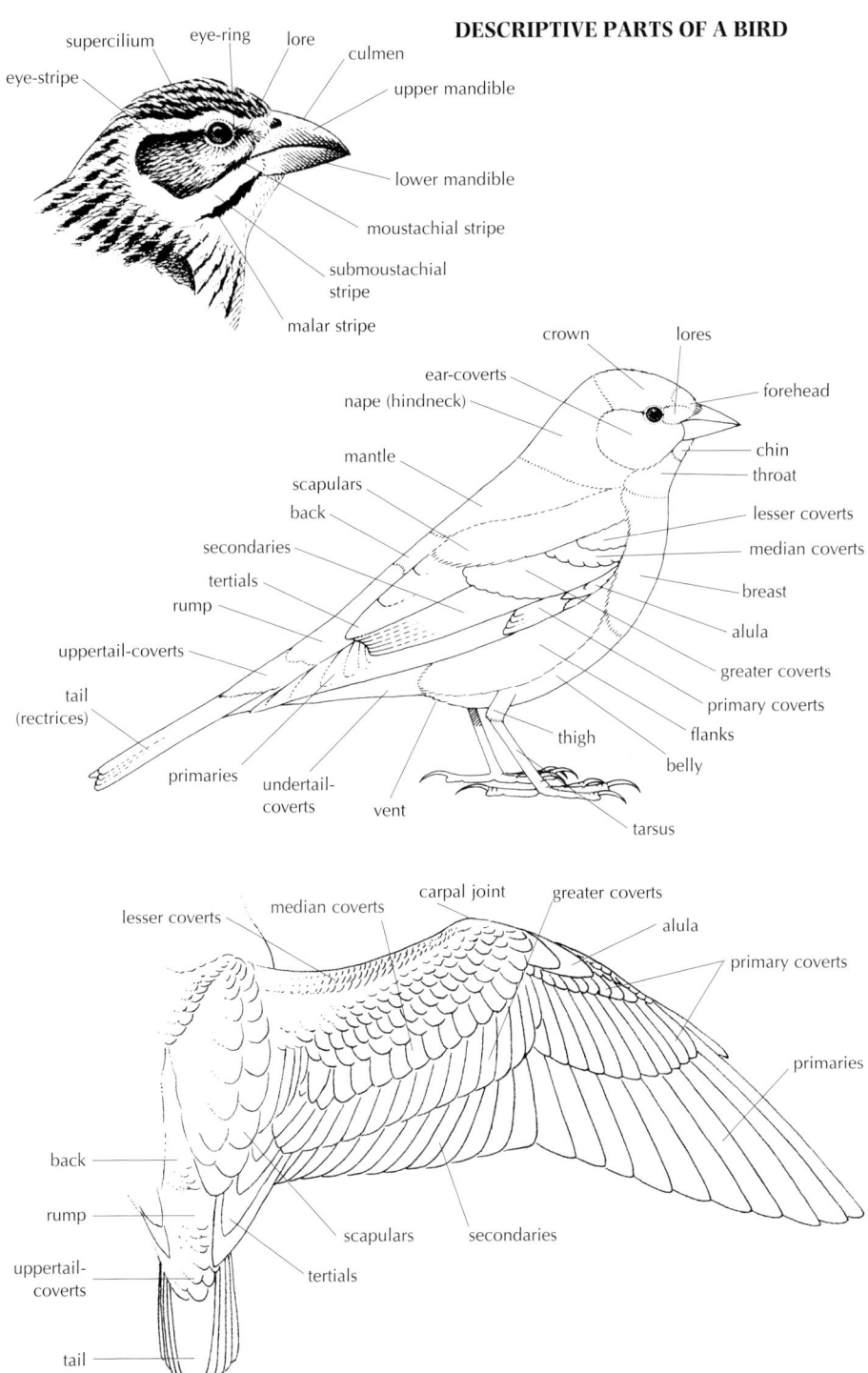

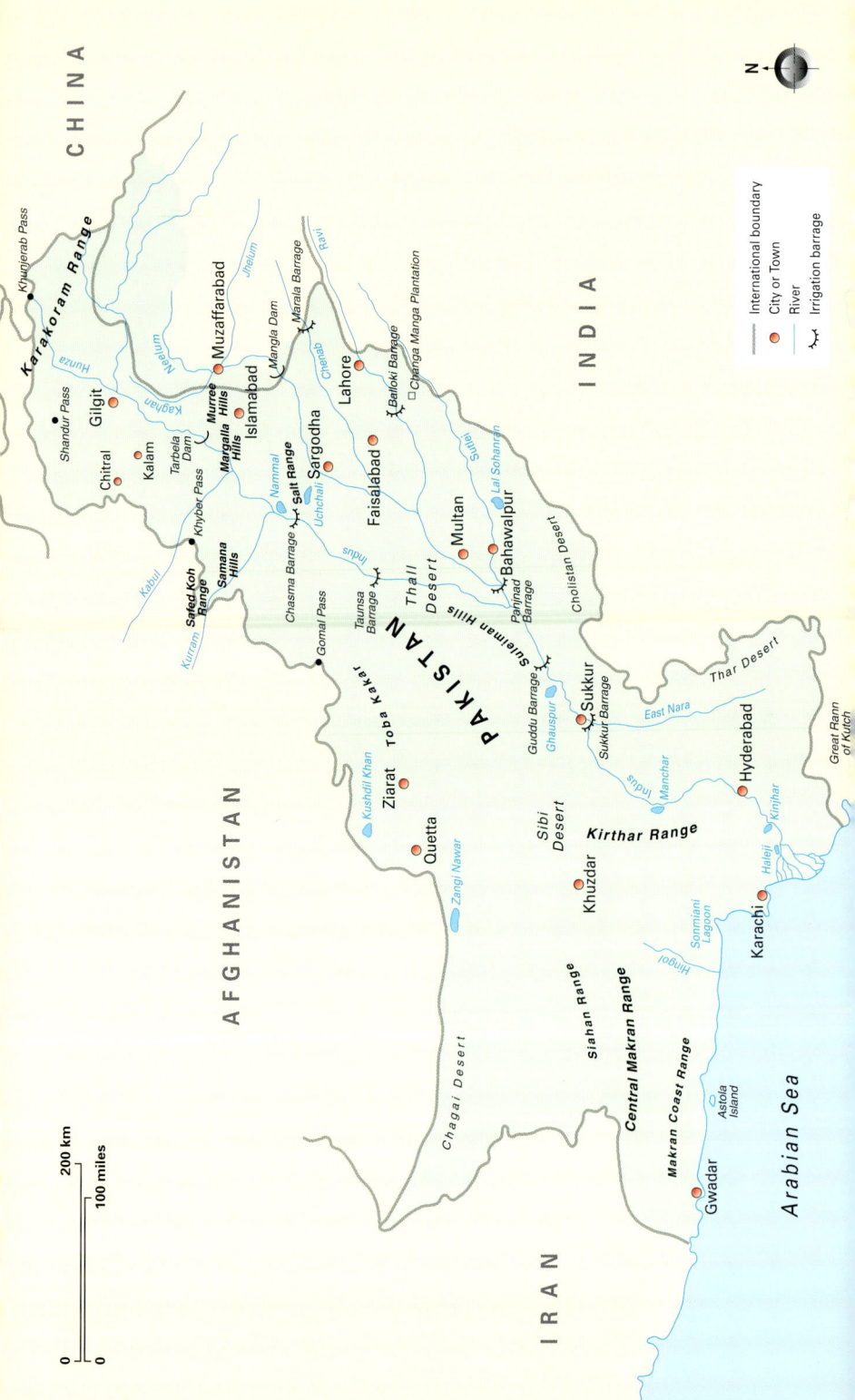

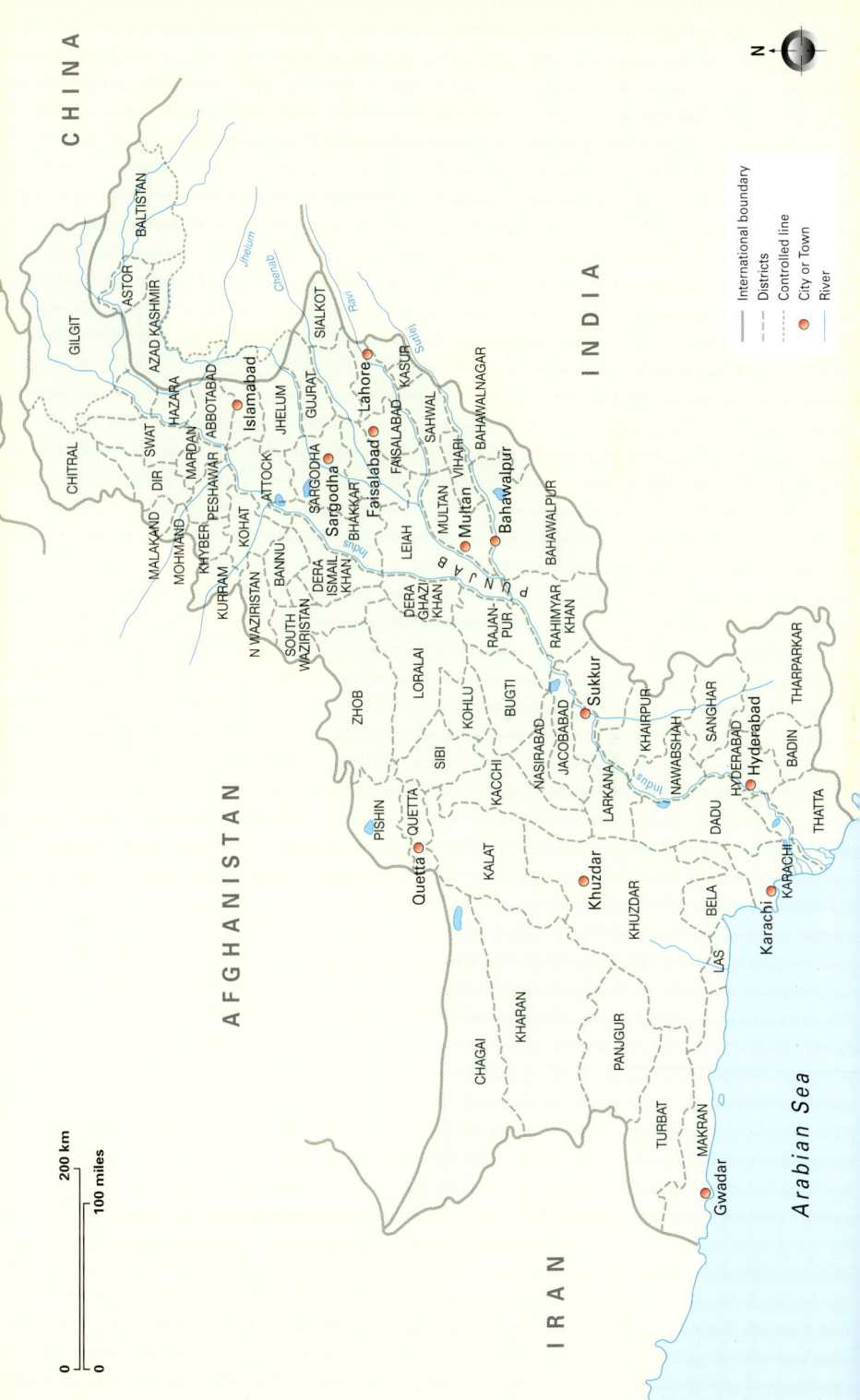

Gyps indicus has been split into two species: the Indian Vulture *G. indicus*, which occurs in Pakistan, and the extralimital Slender-billed Vulture *G. tenuirostris* (Rasmussen and Parry, 2001). The Indian Spotted Eagle *Aquila hastata* has been split from the Lesser Spotted Eagle *A. pomarina*, with the latter being extralimital (Parry *et al.*, 2002). The Golden-spectacled Warbler *Seicercus burkii* is split into several species, including Whistler's Warbler *S. whistleri*, which occurs in Pakistan (Alström and Olsson, 1999). A major new work on the birds of the subcontinent (Rasmussen and Anderton, 2005) has made many novel taxonomic changes, most of which are only discussed briefly. Although some of these changes have been the subject of detailed peer-reviewed papers and have been accepted by other taxonomic authorities, it was decided that, for the purposes of this book, it was best to retain the taxonomy as described above and defer adoption of the new splits and lumps until they had been properly analysed and reviewed.

Colour Plates and Plate Captions

Species that occur regularly in Pakistan are illustrated in colour and described in the plate captions. Vagrants and very rare species, as well as those species now feared extinct, are described in the Appendix, with reference to features distinguishing them from other more regularly recorded species where appropriate. The illustrations show distinctive sexual and racial variation whenever possible, as well as immature plumages. Some distinctive races as well as immature plumages are also depicted. Where possible, species depicted on any one plate have been shown to approximately the same scale.

The captions identify the figures illustrated, very briefly summarise the species' distribution, status, altitudinal range and habitats, and provide information on the most important identification characters, including voice where this is an important or interesting feature. The approximate body length of the species, including bill and tail, is given in centimetres. Length is expressed as a range when there is marked variation within the species (e.g. as a result of sexual dimorphism or racial differences).

The identification texts are based on *Birds of the Indian Subcontinent* (Grimmett *et al.*, 1998). The vast majority of the illustrations have been taken from the same work and wherever possible the correct races for Pakistan have been depicted. The text and plates are based on extensive reference to museum specimens combined with considerable work in the field. The texts covering status, distribution, habits and habitat, as well as the maps, are based on *The Birds of Pakistan* (Roberts, 1991–1992). Globally threatened species (species at risk of global extinction) are indicated as such, following BirdLife International (see www.birdlife.org/datazone/species), with the IUCN threat category given in parentheses.

Key to Maps

The maps found in this book use the following colours to represent areas of Pakistan in which a species can be seen at different times of the year.

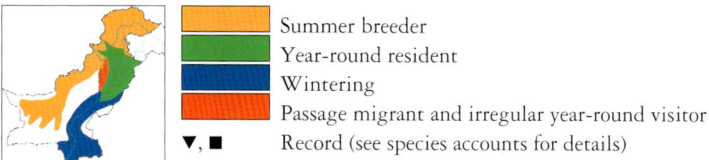

Summer breeder
Year-round resident
Wintering
Passage migrant and irregular year-round visitor
▼, ■   Record (see species accounts for details)

Plumage Terminology

The figures on page 7 illustrate the main plumage tracts and bare-part features. This terminology for bird topography has been used in the plate captions. Other terms that have been used are defined in the Glossary (see pp.29–31). Juvenile plumage is the first plumage on fledging, and in many species it is looser and more fluffy than subsequent plumages. In some families, juvenile plumage is retained only briefly after leaving the nest (e.g. pigeons), or hardly differs from adult plumage (e.g. many babblers), while in other groups it may be retained for the duration of long migrations or for many months (e.g. many waders). In some species (e.g. *Aquila* eagles), it may be several years before all juvenile feathers are finally moulted. The relevance of the juvenile plumage to field identification therefore varies considerably. Some species reach adult plumage after their first post-juvenile moult (e.g. larks), whereas others go though a series of immature plumages. The term 'immature' has been employed more generally to denote plumages other than adult, and is used either where a more exact terminology has not been possible or where more precision would give rise to unnecessary complexity. Terms such as 'first-winter' (resulting from a partial moult from juvenile plumage) or 'first-summer' (plumage acquired prior to the breeding season of the year after hatching) have, however, been used where it was felt that this would be useful.

Many species assume a more colourful breeding plumage, which is often more striking in the male compared with the female. This can be realised either through a partial (or in some species complete) body moult (e.g. waders), or results from the wearing-off of pale or dark feather fringes (e.g. redstarts and buntings).

GEOGRAPHICAL SETTING

Climate

Pakistan covers an area of 796,096 km^2 and stretches from 24°N on the Arabian Sea coast to 37°N, where the northern mountainous regions border the Pamir Wakhan. Most of the country could be described as semi-desert with a subtropical climate. In the major desert tracts, average annual precipitation is only 20–40 mm, with large areas often receiving no rain at all in some years. The highest rainfall zone lies along the Himalayan outer foothills, of which Hazara District and the Murree Hill range experience an average annual precipitation of up to 1350 mm, which in the Murree Hills includes an average of 6.5 m of winter snow.

In the south, some desert tracts suffer an ambient daytime temperature of more than 49°C during the summer months from May to July, whilst in the arid mountainous region west of the Indus, winter temperatures can drop to minus 20°C, and rise to 38°C in summer.

The southwest monsoon is relatively weak by the time it hits the Arabian Sea coastline of Pakistan, and it tends to be deflected eastwards along that side of the Indus river, curving westwards again in the north as it comes up against the Himalayan mountain barrier.

Despite relatively unfavourable climatic conditions for plant and animal life, the region has been greatly influenced by the Indus river. Flowing for over 2896 km within Pakistan, it has experienced a great seasonal variation in its flow levels over the millennia, with water in winter shrinking to less than 50 times the high flow at the end of summer that results from snow melt and monsoon rains. This has enabled some of man's earliest civilisations to colonise the river borders and use trapped flood waters to grow cereal crops.

Cool winter temperatures also present more favourable conditions for an influx of both central Asian bird migrants as well as Himalayan altitudinal migrants.

MAIN HABITATS

These can be roughly divided into four main types.

Indus Basin

Most of the Indus basin and its five tributaries form a vast flat alluvial plain that falls barely 19 cm per km and extends 1126 km from the north to the sea coast in the south. Its composition of fine silt to a depth of up to 150 m has facilitated manual excavation of a huge network of irrigation canals, estimated to extend for over 56,315 km. All these major canals, branches and distributaries have roadside tree plantations and grass-covered embankments in an otherwise arid desert environment, providing significant habitat for birds. Original riverine forest is now reduced to a few relict patches. Original swamps have also largely gone, but wetland habitats remain in the form of seepage areas, which create a significant micro-habitat in an otherwise desert environment and occur all along the sides of major canals and, especially, irrigation barrage headponds. Further wetland habitat is provided by huge reservoirs such as the Mangla and Tarbela dams on the Indus and Jhelum, which store water from the summer snow melt and monsoon.

The former habitat in this region was tropical thorn forest, but this has largely been cleared for cultivation over the past 1000 years, and only small pockets remain (sometimes recently regenerated). Areas of mangrove occur at the coast, especially in the Indus delta, where there are extensive intertidal mud and sand flats.

Arid Mountains and High Stony Plateaux

This comprises most of the area west of the Indus river from the sea coast up to the North West Frontier Province (NWFP). The sheltered slopes and valleys have a stunted tree and bush cover of deciduous acacia, 'Jujub' or 'Ber' (*Zizyphus*), pistachio and Wild Olive scrub. At higher elevations is open-canopy juniper or Chilghoza and Blue Pine forest. This area has lost much of its original tree cover as a result of livestock grazing and browsing, as well as cutting for fuel wood.

PURE DESERT SAND DUNES (CHAGHAI DESERT).

Pure Desert Tracts

There are three main desert regions on Pakistan's borders, and two less extensive ones in the interior. Bordering the great Seistan Desert basin of Iran and Afghanistan is the Chaghai Desert. Bordering Rajasthan in India is the Great Thar Desert and, further north, bordering the former princely state of Bikaner, the Cholistan Desert. West of the ancient riverside town of Sukkur in northern Sind lies the Sibi Desert, with a reputation for being the hottest place on the subcontinent. In May and June, ambient temperatures of 52°C are regularly recorded at Sibi town. In the western part of the Punjab lies the Thal Desert, in many parts now irrigated with pumped groundwater.

Himalayan Mountain Ranges and Foothills

In the north and within Pakistan there is a confluence of five great mountain chains: the Hindu Kush, the Safed Koh, the Pamirs, the Himalayan Pir Panjal and the Karakorams. The Karakorams contain the greatest concentration of high peaks anywhere in the world: 18 are over 6240 m in height. The Himalayan chain averages only 80 km in width, whereas the Karakorams average over 192 km in width.

A complex range of vegetation zones extends from the foothills to the high Himalayas and beyond. At low elevations there is a strip of subtropical dry deciduous scrub forest, with fine regenerating tracts in the Margalla Hills behind Islamabad. Beyond this is a band of subtropical Pine forest dominated by Chir Pine, where the undergrowth is typically limited due to periodic forest fires.

At higher elevations is a band of Himalayan moist temperate forest, which is particularly important for wildlife. It is predominantly a coniferous forest with glades of mixed deciduous broadleaved species. Commercial logging continues to have a major impact in this zone, as does livestock grazing and browsing. This forest is typified by

PURE DESERT (CHOLISTAN DESERT).

HIGH MOUNTAIN RANGES (SKARDU-BALTISTAN).

most of the Murree Hill range and the lower Neelum and Kaghan valleys, and extends westwards to parts of eastern Swat bordering on Indus Kohistan. Above the coniferous forest is White Birch forest and thickets of Wild Rose and berberis, and then alpine meadows, rocks and scree, which are subject to heavy grazing pressure in summer.

Mountain areas that are less subject to monsoon influence are characterised by a drier coniferous forest with Evergreen Oak.

North of the main Himalayan range is an area of cold desert and dry alpine habitat, as found in northern regions of Chitral, Gilgit, Hunza and all of Baltistan. In the valley floor the landscape is boulder-strewn, with vegetation along steams and in areas of moisture below glaciers and snowfields.

IMPORTANT BIRD SPECIES

Restricted-range Species

A significant part of the Western Himalayan Endemic Bird Area (EBA), as designated by BirdLife International, lies within Pakistan territory (Stattersfield *et al.*, 1998). This EBA has 11 restricted-range species, of which ten are known from Pakistan: Western Tragopan, Cheer Pheasant, Brooks's Leaf Warbler, Tytler's Leaf Warbler, Kashmir Flycatcher, White-cheeked Tit, White-throated Tit, Kashmir Nuthatch, Spectacled Finch and Orange Bullfinch. All of these are resident or summer breeders in, or adjacent to, temperate Himalayan forest. This habitat is under extreme pressure from commercial forestry and livestock grazing and browsing. Other species in this region with limited global ranges (but not meeting the restricted-range criteria set by BirdLife International) include Mountain Chiffchaff and Long-billed Bush Warbler, with Plain Leaf Warbler and Red-mantled Rosefinch in the drier mountains along the Afghanistan border.

The Indus plains are designated as a Secondary Area by BirdLife International as a result of the presence of the restricted-range Sind Sparrow. Also present in this region are several other species with limited global ranges, including Rufous-vented Prinia and Jerdon's Babbler.

Globally Threatened Species

BirdLife International considers 26 species occurring in Pakistan as Globally Threatened (see www.birdlife.org/datazone/species/index.html). Most dramatic has been the decline in White-rumped Vulture and Indian Vulture; both species have suffered catastrophic losses from poisoning as a result of the widespread veterinary use of the anti-inflammatory drug diclofenac on draught animals, which after death are scavenged by vultures. It causes renal failure in these birds. It is hoped that this drug can be removed from the environment and that vulture populations will then recover.

Lakes in the Punjab Salt Range support an important wintering population of White-headed Duck. Pakistan's wetlands also attract wintering Imperial Eagle, and the more numerous Greater Spotted Eagle, with the latter occasionally breeding in lower Sind. Pallas's Fish Eagle is resident at a number of wetlands in the Indus basin but has declined throughout its former range. Sociable Lapwing also occurs as an occasional winter visitor in the Indus basin. Seepage areas and remnant marshes along the Indus river support the entire population of Jerdon's Babbler ssp. *scindicum*. Indian Skimmer is a summer visitor to the Indus river, where it breeds, although this species has declined markedly in recent decades. There is a small breeding population of Marbled Duck at Zangi Nawar lake in Baluchistan, and occasionally at one or two other sites, with the species occurring more widely in winter. Dalmatian Pelican is a regular passage migrant and winter visitor. Other Globally Threatened waterbirds occurring in Pakistan are rare visitors or perhaps now extinct in the country.

Arid regions of the Thar Desert adjoining India still hold Indian Bustard and perhaps Stoliczka's Bushchat, although presence is affected by the strength of the monsoon. In good monsoon years, Lesser Florican may still arrive to breed in grasslands in border areas of Sind and Punjab. The temperate Himalayan forests support important populations of Western Tragopan, with the Palas Valley in Indus Kohistan being perhaps the most important location in the world for this spectacular species. Two further Himalayan species, Cheer Pheasant and Kashmir Flycatcher, occur in Pakistan, but their status is poorly known and the Cheer Pheasant may now be extinct in the country.

Macqueen's Bustard (treated as conspecific with Houbara Bustard by BirdLife International) is a regular winter visitor, with attempted breeding in southwest Baluchistan during the 1990s. The species has suffered heavy predation from Arab hunting parties, who regard it as the special quarry of their falcons. The practice of falconry has put considerable pressure on falcon populations, too, with birds being widely trapped in Pakistan, and Saker Falcon has recently been listed as Globally Threatened.

MIGRATION

Pakistan lies at a crossroads for bird migration as a result of its strategic geographical location. The Indus basin is one of the world's great migratory flyways, with migrant birds seeking to avoid the highest mountain ranges in the north and the vast dry deserts to the west and east, and is the principal route followed by many species that breed extralimitally and winter in Pakistan or in other regions of the Indian subcontinent. This includes a wide variety of ducks and waders, raptors, and passerines such as warblers, pipits and buntings. The spring and autumn passage of Red-throated Flycatcher, Rosy Starling, and Red-headed and Black-headed buntings, which winter largely in India, is particularly conspicuous.

Some species enter the subcontinent via northern tributaries of the Indus (such as the Gilgit and Hunza rivers). Others, including many Common and Demoiselle cranes, snipe and pelicans, enter from further west, using the Peiwar Pass, and therefore bypassing the Safed Koh mountain range and entering via the Kurram Agency of North West Frontier Province.

The rich feeding conditions along the Rift Valley of eastern Africa are the winter destination for a number of species that breed in central Asia and Siberia, and some of these species, such as the Greater Whitethroat, Rufous-tailed Rock Thrush and Red-backed Shrike, migrate there in autumn through the more hilly regions of western Pakistan. On their return migration in the spring, these species choose a more westerly route through neighbouring Iran and Afghanistan. A number of species that breed in Pakistan, such as the Blue-cheeked Bee-eater, European Roller and Eurasian Nightjar, also head for Africa in winter.

Favourable feeding conditions brought on by the summer monsoon also bring a small number of purely Oriental species from India and perhaps further east, as summer migrant breeders. These include the Yellow-wattled Lapwing, Watercock, Indian Skimmer and Common Hawk Cuckoo. The temperate forests in northern Pakistan are a breeding destination for species that winter in India, such as the Rusty-tailed Flycatcher, Indian Blue Robin, and Tytler's and Western Crowned leaf warblers. A majority of species breeding in the Himalayas undergo some form of limited altitudinal migration to avoid the harsh winter conditions, either descending to lower elevations and south-facing slopes, or to the plains.

BIRDWATCHING AREAS

Only a selection of sites is described here. A variety of birds can also be found in urban parks and older gardens, as well as in intensively farmed areas.

Karachi Area

Buleji Beach and Cape Monze
Location: On the outskirts of Karachi city along the western seaboard. Buleji is about 39 km from the city centre and Cape Monze about 45 km.
Habitat: Rocky coastline with intervening sandy beaches, and some degraded mangrove at the southern end.

SEEPAGE ZONE BY IRRIGATION BARRAGE (JINNAH BARRAGE ON THE INDUS).

Birds: Pallas's, Sooty, Brown-headed and Slender-billed gulls, and Caspian, Gull-billed, Great Crested, Lesser Crested and Saunders's terns occur throughout the year. During the monsoon, White-cheeked Tern and Wilson's Storm-petrel can be seen offshore.

Ghizri Creek
Location: On the southern outskirts of Karachi, about 12 km from the city centre.
Habitat: Tidal creek with mudflats and limited salt marsh, surrounded by sand dunes.
Birds: In winter, Curlew and Broad-billed sandpipers, Dunlin, Pacific Golden Plover, Pied Avocet, Greater and Lesser sand plovers, Western Reef Egret, Saunders's Tern and Long-legged Buzzard occur.

Haleji and Hadiero
Location: Haleji Reservoir is about 60 km west of Karachi on the old highway to Hyderabad city. Hadiero is further north on a branch road to Jamshoro and about 73 km from Karachi.
Habitat: Haleji has a tree-lined embankment (suitable to drive on), with seepage pools and wetlands along its fringes, dense patches of *Typha* and lotus lilies, and a steep rocky island towards the north end.
Birds: A large population of Purple Swamphen is resident. At least 13 species of duck, as well as Dalmatian and Great White pelicans, can be seen in winter, as can a variety of raptors, including Osprey, Greater Spotted and Imperial eagles, and Eurasian Marsh Harrier. Breeding raptors include Brahminy Kite, Pallas's Fish Eagle and Red-necked Falcon. In summer, 13 species of herons and bitterns can be seen, and in the seepage pools in winter, there are nesting Cotton Pygmy-goose, Purple Heron, Glossy Ibis, Eurasian Spoonbill and Streaked Weaver. Hadiero, being a larger water body, is a good place in winter for raptors: Osprey, White-tailed Eagle, Cinereous Vulture, Eurasian Griffon, Short-toed Snake Eagle, Eurasian Marsh Harrier and Greater Spotted Eagle. In the adjoining thorn scrub, Orphean and Desert warblers occur in winter, and Graceful Prinia, Desert Lark, and Ashy-crowned and Black-crowned sparrow larks occur year-round.

Hab Dam and Valley
Location: 30 km north of Karachi city.
Habitat: The reservoir is flanked by desert scrub, with some flash-flood gullies and higher rocky hills on its western side.
Birds: A good place to see Hume's Wheatear, Pale Crag Martin, White-eared Bulbul, Grey Hypocolius and House Bunting, as well as Painted and Chestnut-bellied sandgrouse and Cream-coloured Courser. On the reservoir in winter, there are Black-necked Grebe, Common Merganser and diving ducks.

Upper Sind

Kandkhot and Ghauspur
Location: 109 km northeast of Sukkur.
Habitat: A series of flood-protection embankments and seepage swamps or temporary pools, plus the permanent Ghauspur Lake. There is a rest house at Kandkhot.
Birds: A good place to see breeding Pallas's Fish and Tawny eagles, and Sykes's Nightjar. In winter, Yellow-eyed Pigeon, White-tailed Eagle, Black-bellied Tern and many waders, especially Black-tailed Godwit, occur. Fifteen species of raptor, six species of terns and seven species of herons/bitterns have been recorded.

SEEPAGE ZONE BY HALEJI RESERVOIR, SIND.

Hingol National Park
Location: Though it is in Baluchistan at the mouth of the Hingol river, the park is best reached from Karachi. It lies approximately 252 km west of Karachi on the Arabian Sea coast.
Habitat: Arid rocky hills down to the sea, with the Hingol river providing some scattered Oleander and deciduous tamarisk. On the surrounding hills grow *Artemisia* and the dwarf Mazri Palm.
Birds: Pallid Swift, Black-bellied Sandgrouse, Sooty Falcon, Brown-necked Raven, Hooded Wheatear and Chestnut-shouldered Petronia, and, in winter, Dalmatian and Great White pelicans, many waders, Laggar Falcon, Eurasian Hobby and Finsch's Wheatear.

Quetta Area, Baluchistan Province

Hazar Ganji National Park
Location: 20 km southwest of Quetta.
Habitat: The Chiltan range of hills, with the highest peak at 3264 m, falling steeply to stony screes and slopes at the eastern end of the park at 2000 m. Dotted with juniper at higher elevations, and at their foot with scattered pistachio and Mountain Ash (*Fraxinus xanthoxyloides*), *Artemisia* and dramatic yellow globes of *Ferula oopoda*.
Birds: Breeding See-see and Chukar partridges, European Bee-eater, Pallid Scops Owl, Eurasian Nightjar, Red-rumped Swallow, Bay-backed Shrike and Desert Finch. At higher elevations there is a small population of Yellow-billed Chough.

Ziarat
Location: 96 km west of Quetta.
Habitat: This is a high-altitude valley, situated at about 2400 m, with the highest nearby peak Mount Kaliphat at 3945 m. A climax habitat, exceptionally restricted on a worldwide basis, it comprises open-canopy tall juniper forest with flowering tulips, Foxtail Lilies and purple tussocks of *Onobrychis* in summer. There is some deciduous fruit cultivation in the valley.
Birds: Summer breeding Eurasian Scops Owl, Eurasian Sparrowhawk, Mistle Thrush, Spotted Flycatcher, Black Redstart, Lesser Whitethroat (ssp. *althaea*), Eastern Rock

Nuthatch, Eurasian Crag Martin, Sulphur-bellied and Plain leaf warblers, Fire-fronted Serin, White-winged Grosbeak, Common and Red-mantled rosefinches and White-capped Bunting.

Torghar Wildlife Preserve
Location: 52 km west of the town of Zhob, which is about 334 km northwest of Quetta.
Habitat: Bush-dotted mountainous country with scattered Wild Olive, pistachio and Mountain Ash (*Fraxinus xanthoxyloides*), *Sophora* and *Ephedra* bushes, and Foxtail Lilies adding colour to the scene in summer.
Birds: Breeding Sind Woodpecker, Eurasian Nightjar, Lammergeier, Cinereous Vulture, Eurasian Griffon, Booted Eagle, Blue Rock Thrush, Spotted Flycatcher, Black Redstart, Orphean Warbler, Sulphur-bellied and Greenish (ssp. *nitidus*) warblers, White-cheeked Tit, Common Rosefinch, Fire-fronted Serin and Hawfinch, and, in winter, Dark-throated Thrush, Brambling and Chaffinch.

Punjab

Lal Sohanran Desert Park
Location: 26 km east of Bahawalpur city.
Habitat: A series of tamarisk-fringed lakes where surplus summer canal water is stored, bordered to the south by the Cholistan Desert, with much lopped *Prosopis*, acacia trees, tamarisk and *Calligonum* bushes.
Birds: In the wetlands, there are many species of duck, including Smew in winter, grebes (including Horned), Darter, all three species of cormorant, and Black Stork and Demoiselle Crane on passage, and in winter there are many raptors. Breeding birds include Black-crowned Night Heron and Pallas's Fish Eagle. In the desert, there are Sykes's Nightjar, Spotted, Chestnut-bellied and Black-bellied sandgrouse, Macqueen's Bustard and Greater Hoopoe Lark, and in the adjacent forest plantation is a winter roost of Long-eared Owl.

TORGHAR WILDLIFE PRESERVE, ZHOB DISTRICT, BALUCHISTAN.

Kal Chitta Hills–Punjab Salt Range
Location: 75 km west of Islamabad.
Habitat: Low rocky hills with a good ground cover of stunted Wild Olive, acacia, and bushes of *Adhatoda* ('Baikar') and *Dodonaea* ('Sanatha').
Birds: In winter, there are Alpine Swift, Common Kestrel, Blue Whistling Thrush, Rufous-backed and Black redstarts, Lesser Whitethroat (ssp. *minula*), Orange-flanked Bush Robin, Variable Wheatear, Great Tit, Hawfinch and Rock Bunting. Resident are Grey Francolin and See-see Partridge, Sind Woodpecker, Common Wood Pigeon, Short-toed Snake and Bonelli's eagles, Small Minivet, Brown Rock-chat and Pale Martin.

Uchchali and Khabbaki Complex (Salt Range Lakes)
Location: From Sargodha town on the plains, approximately 99 km to Khabbaki Lake and 130 km further west to Uchchali Lake.
Habitat: Brackish lakes, surrounded by scrub-covered hills.
Birds: The best place to see wintering White-headed Duck. Also here are wintering diving duck, Great Crested, Black-necked and Little grebes, and, on Uchchali, Greater Flamingos. In the surrounding area are Buff-bellied and Long-billed pipits, and Grey-necked Bunting.

Lahore Area

Balloki Headworks
Location: Irrigation barrage on the Ravi river, 69 km southwest of Lahore.
Habitat: Seepage swamps alongside flood-protection embankments.
Birds: Great Bittern, Spotted and Baillon's crakes, White-tailed Stonechat and Striated Grassbird. In spring, four subspecies of Yellow Wagtail occur in full breeding plumage. Savanna Nightjar and Watercock are present in summer.

Changa Manga Forest Plantation
Location: 86 km southwest of Lahore.
Habitat: Reputedly the oldest irrigated forest plantation on the subcontinent, planted in the late 19th century to provide timber for railway sleepers. Planted today with *Dalbergia* ('Shisham') and white mulberry for manufacturing sports goods.
Birds: Indian Grey Hornbill, Yellow-footed Green Pigeon, White-browed Fantail, Oriental Magpie Robin, Ashy Prinia, Asian Pied Starling and, in summer, Pied and Common Hawk cuckoos and Eurasian Golden Oriole. In winter, Bluethroat, Red-throated Flycatcher and Black Redstart are seen. In spring, there is a major passage of Rosy Starling and Common Rosefinch.

Islamabad Area

Rawal Lake
Location: East of the city outskirts and adjacent to the old Murree road, about 0.8 km from the city centre.
Habitat: Western edges of the lake are reed-fringed with *Typha* and *Phragmites* and surrounded by scattered trees, mostly acacia ('Babul' or 'Kikar'), mango and *Bauhinia*. The exotic paper mulberry (*Broussetia*) is rampant everywhere.
Birds: The lake attracts many spring and autumn migrants, including Smew, Common Merganser, storks and waders. In spring, Great Crested Grebes have been seen giving

full courtship display, and this is a good place to see 'White-headed' Yellow Wagtail *M. flava leucocephala*. Crested Kingfisher occurs on some of the west-bank feeder streams.

Margalla Hills
Location: A national park, lying on a north–south axis along the western boundary of the city, accessible on foot from the northwestern urban area.
Habitat: Rising from about 500 m at the base to 1600 m on the highest summit ridges, these hills are arguably the most diverse area in Pakistan. They have a good mixture of Indo-Malayan plants, mostly rather stunted, but with thick undergrowth of *Carissa* ('Garunda'), *Adhatoda* ('Baikar') and *Zizyphus* ('Ber'). Taller trees include indigenous Silk Cotton, *Ficus* spp., Indian Laburnum, and various acacias.
Birds: In winter, there is a chance to see many Himalayan altitudinal migrants, including Rusty-cheeked Scimitar Babbler, Black-chinned Babbler and Red-billed Leiothrix, as well as Grey Treepie and White-throated Fantail. The Asian Study Group, a society strongly supported by ex-patriot diplomats, listed 356 species in the 1990s from this area, and two more have since been recorded around Rawal Lake.

Outer Murree Foothills
Location: Lying to the north of Islamabad, varying from 20 km to 32 km in such directions as the Tret Valley, Kahuta and Lehtrar.
Habitat: Lying mainly between 900 m and 1500 m. Predominantly subtropical long-leaved pine ('Chir'), with a good mix of deciduous trees in sheltered ravines, including pistachio, Silver Oak ('Ban') and an understorey of barberry, wild pomegranate and *Zizyphus*.
Birds: A good place to see Brown-fronted and Grey-capped Pygmy woodpeckers, Rosy Minivet, Chestnut-bellied Nuthatch, Thick-billed Flowerpecker and nesting Plum-headed Parakeet.

HIMALAYAN COOL TEMPERATE FOREST: MIXED DECIDUOUS AND CONIFERS (DUNGA GALI, MURREE HILLS).

Hazara District, North West Frontier Province

The Murree Hills and Galis
Location: Nathia Gali, 42 km north of Murree town, and at 2400 m elevation, has many tourist hotels and is the best place to survey the area.
Habitat: On warmer south-facing slopes, Blue Pine and Deodar forest, and on northern slopes, Silver Fir, and a mix of deciduous trees such as Himalayan Horse Chestnut, elm and Bird Cherry. The understorey includes *Viburnum*, Bush Honeysuckle and attractive climbers such as the White Rose and Mountain Clematis.
Birds: Koklass Pheasant, Common and Indian cuckoos, Mountain Scops Owl and Collared Owlet, Grey Nightjar, Yellow-billed Blue Magpie, Eurasian Jay, Chestnut-bellied and Blue-capped rock thrushes, Grey-winged Blackbird, Chestnut Thrush, Rufous-bellied Niltava, Verditer Flycatcher, Ultramarine and Dark-sided flycatchers, Indian Blue Robin, Orange-flanked Bush Robin, and Lemon-rumped, Hume's and Western Crowned warblers.

Kaghan Valley
Running due north from the town of Balakot, this is actually the Kunhar river. It is a scenic and popular tourist area. Four localities along the valley offer interesting birdlife.

– Balakot
Location: At about 900 m elevation, where the valley starts to close in.
Habitat: The surrounding hills are mostly terraced for cultivation, with some scattered long-needle pine ('Chir') and Silver Oak ('Ban') in side ravines. The terraced embankments are clothed with *Indigofera* and *Reinwardtia* ('Basant').
Birds: Eurasian Jackdaw nests in colonies in earth cliffs or old trees, Crested and White-capped buntings on the open slopes, and Asian Paradise-flycatcher in the shaded ravines.

– Shogran and Shahran
Location: Shogran lies on the east bank of the river, 54 km from Balakot, and Shahran on the west bank, 62 km from Balakot. Both are accessible via a 4WD track from Paras.
Habitat: Located at approximately 2700 m in open valleys with Silver Fir and Deodar forest, and some deciduous trees such as Hill Oak ('Moru'), Bird Cherry, Horse Chestnut and walnut. Good trails lead to alpine meadows above 3200 m.
Birds: Koklass Pheasant, Himalayan Monal, Eurasian Woodcock, Common Kestrel, Scaly Thrush, White-bellied Redstart, Grey-sided Bush Warbler, Goldcrest, White-throated Tit, Rosy Pipit and Orange Bullfinch. There is an excellent variety of breeding *Phylloscopus* warblers, including Large-billed, Tytler's, Western Crowned, Lemon-rumped, Tickell's, Brooks's Leaf and Hume's.

– Naran
Location: This is a popular resort with limited hotel accommodation, 98 km north of Balakot, with a side valley leading to an alpine lake, Saif-ul-Maluk.
Habitat: Varying, from the main river with lopped mulberry, stands of Silver Fir, Deodar and Blue Pine, with thickets of barberry and dwarf elder, and ascending to birch scrub with dwarf juniper and stony screes.
Birds: Stream-side birds include White-capped Water Redstart, Brown Dipper and Citrine Wagtail. At higher elevations, Himalayan Snowcock, Snow Pigeon, Lammergeier, Eurasian Blackbird, Blue-fronted Redstart and White-tailed Rubythroat occur.

– Battakundi
Location: Approximately 122 km north of Balakot.
Habitat: At the upper limit of the tree-line, with slopes covered by dwarf elder, creeping juniper, and open meadows.
Birds: Himalayan Snowcock, Snow Pigeon, Lammergeier, Himalayan Griffon, Golden Eagle, Common Kestrel, Eurasian Blackbird, White-throated Dipper, Yellow-billed Chough, Wallcreeper, Pale Martin, Oriental Skylark, Rosy Pipit, Rufous-breasted Accentor, Plain Mountain Finch, Spectacled Finch and Eurasian Goldfinch.

Northern Areas

Swat
Bounded to the north by Gilgit and to the east by the Indus river, Swat lies north of Nowshera District in North West Frontier Province. The capital, Saidu Sharif, lies in a broad valley at about 1500 m, with rice cultivation in summer, and roads planted with Persian Lilac and Oriental Plane trees.

Kalam
Location: This hill resort lies at an elevation of 2400 m, and approximately 80 km north of Saidu Sharif. There are several good tourist hotels.
Habitat: Pure stands of Deodar forest with Holly Oak on the fringes and along river banks. Silver Fir on the higher slopes.
Birds: Oriental Turtle Dove, Tawny and Eurasian scops owls, Black-billed Magpie, Yellow-billed Blue Magpie, Blue-capped and Blue rock thrushes, Chestnut, Tickell's, Mistle and Blue Whistling thrushes, Rusty-tailed Flycatcher, White-capped and Plumbeous water redstarts, and Brooks's Leaf, Hume's, Western Crowned and Lemon-rumped warblers.

Gilgit
Accessible via the spectacular Karakoram Highway, or by daily flights (weather permitting) from Islamabad.

KAGHAN VALLEY: HAZARA DISTRICT – ALPINE VALLEY ABOVE BATTAKUNDI.

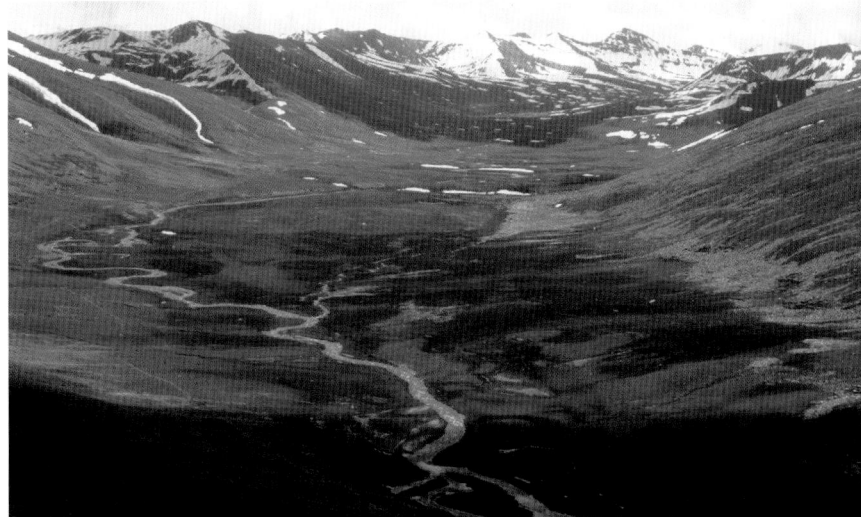

Town Environs

Location: Gilgit river valley at 1200 m, where it can be hot in summer. Winter temperatures go down to minus 12°C, but its dry atmosphere makes it pleasant by day.
Habitat: Bordering the Gilgit river, with scattered apricot orchards, and cereal cultivation in summer.
Birds: In winter, feeding in stubble fields are Rock Sparrow and Pine and Rock buntings, and by the river are Solitary Snipe, Northern Lapwing, Common Sandpiper and Rufous-backed Redstart. In the town, in winter, are Black-billed Magpie, Winter Wren and Wallcreeper. In summer, European Bee-eater, Alpine Swift and Eurasian Golden Oriole are seen.

Naltar Valley

Location: At 3300 m elevation, this forested valley is located approximately 58 km from Gilgit town, on the west bank of the Hunza river.
Habitat: Mixed forest of Silver Fir and Blue Pine, and with terraced cultivation, and scree slopes with Wild Rose, Wild Gooseberry and creeping juniper.
Birds: Eurasian Eagle Owl, Tawny Owl, mixed flocks of Red-billed and Yellow-billed choughs, Spotted Nutcracker, Brown Dipper, Mistle Thrush, Blue-capped Redstart, Little Forktail, Long-billed Bush Warbler, White-browed Tit Warbler, Rufous-breasted Accentor and White-browed Rosefinch. A variety of breeding *Phylloscopus* warblers, including Sulphur-bellied, Western Crowned, Hume's, Lemon-rumped, Brooks's Leaf and Tickell's, and Mountain Chiffchaff.

Hunza Valley, up to the Khunjerab National Park

The Karakoram Highway leads over the Khunjerab Pass at 4780 m into Chinese Xinjiang. There are tourist hotels at intervals along this highway, the last one at Sost, the border and customs post.
Location: Sost is 257 km from Gilgit town; the top of the Khunjerab Pass is 83 km from Sost.
Habitat: Alpine tundra, with many species of *Primula* and *Potentilla*. Lower down the valley, tamarisk and willow grow along the river banks, and there are apricot orchards and poplars in the villages.
Birds: Golden Eagle, Himalayan Griffon, Common Kestrel and, on passage, Lesser Kestrel. Snow Partridge, Himalayan Snowcock, Chukar, Common Swift, Blue Rock Thrush, Black Redstart, White-winged Redstart, Pied Wheatear, Eurasian Crag Martin, Horned Lark, Hume's Short-toed Lark, Alpine Accentor, Brandt's Mountain Finch, Twite, Mongolian Finch, and Red-fronted and Great rosefinches. Northern Wheatear occurs on passage.

Shandur Plateau

Location: Approached from Chitral via Mastuj village, or from Gilgit up the Yasin river valley, on good 4WD roads. It is about 203 km from Gilgit town.
Habitat: A wide grassy valley with a small lake and scree-covered slopes, at an elevation of 3660 m.
Birds: Himalayan Snowcock, Lammergeier, Himalayan Griffon, Golden Eagle, Red-billed and Yellow-billed choughs, Black Redstart, White-winged Redstart, Pied Wheatear, Horned Lark, Oriental Skylark, Rufous-breasted Accentor, Citrine Wagtail, Twite and Great Rosefinch.

NATIONAL ORGANISATIONS

Ornithological Society of Pakistan (OSP)
P.O. Box 73, 109/D, Dera Ghazi Khan, 32200, Pakistan.
email: osp@mul.paknet.com.pk
The mission of the OSP is to conserve and help the sustainable use of the natural resources of Pakistan with particular reference to birds and their habitats. Key activities are to:

- provide advocacy and support for scientific research for species and ecosystem management;
- raise environmental awareness in general and on birds in particular;
- maintain national database on birds of Pakistan, with over 150 years of references;
- work on the first National Red Data Book of birds of Pakistan;
- assist other working groups, NGOs, governmental organisations and universities with conservation issues and priorities in Pakistan.

World Wide Fund for Nature (WWF) Pakistan
P.O. Box 5180, Ferozepur Road, Lahore.
email: info@wwf.org.pk
Established in 1970, WWF-Pakistan is the largest non-governmental conservation organisation in the country. WWF-Pakistan is committed to saving wildlife species and their habitats and to the promotion of nature conservation and environmental protection for sustainable development. WWF-Pakistan's staff of over 250 people is based in the Lahore headquarters, and includes a network of six regional offices in Gilgit, Islamabad, Karachi, Muzaffarabad, Peshawar and Quetta, and 18 project offices (a few examples include Ayubia, Chitral, Jhangar, Sonmiani and Sukkur).

WWF-Pakistan is one of the fastest-growing offices of the worldwide WWF network, which includes 30 National Organisations and Associates, 24 Programme Offices, and a coordinating International Secretariat located in Gland, Switzerland. It is supported by over 5.3 million individual members all over the world. WWF's mission is to stop the degradation of the planet's natural environment and to build a future in which humans live in harmony with nature, by:

- conserving the world's biological diversity;
- ensuring that the use of renewable natural resources is sustainable;
- promoting the reduction of pollution and wasteful consumption.

Punjab Wildlife Research Institute
Wildlife census, Gatwala, Faisalabad.
Tel: 0092 41 9230171
Carries out baseline surveys of protected and unprotected areas, including bird surveys, as well as studies on the interaction between agriculture and birds. One of the key objectives of the institute is also to provide training to staff of the Punjab Wildlife and Parks Department.

Zoological Survey Department, Government of Pakistan
Block No. 61, Pakistan Secretariat, Sharah-e-Iraq, Karachi-74200.
The Zoological Survey Department has undertaken many ornithological surveys, especially on the Rann of Kutch border, where it proved Lesser Flamingo breeding.

Sindh Wildlife Department
Sindh Centre Building, opposite PIA Booking Office, MD Wafai Road, Karachi.
Tel: 9204951-2
Fax: 9204959
The Sindh Wildlife Department has carried out surveys in Zhob, Baluchistan, the Deosai plateaux and other important places. The department has played a key role in protecting high-profile areas of Sind, including Kirthar National Park, Indus Dolphin Reserve between Guddu and Sukkur barrages, Nara Desert Wildlife Sanctuary and marine turtle nesting beaches in Sind.

REGIONAL AND INTERNATIONAL ORGANISATIONS

BirdLife International
Wellbrook Court, Girton Road, Cambridge, CB3 0NA, UK.
Website: www.birdlife.org
Publication: *World Birdwatch* magazine (quarterly).
BirdLife International is a global partnership of conservation organisations that strives to conserve birds, their habitats and global biodiversity, working with people towards sustainability in the use of natural resources. BirdLife Partners operate in more than 100 countries and territories worldwide. BirdLife's aims are to:

- prevent the extinction of any bird species;
- maintain and, where possible, improve the conservation status of all bird species;
- conserve and, where appropriate, improve and enlarge sites and habitats important for birds;
- help, through birds, to conserve biodiversity and to improve the quality of people's lives;
- integrate bird conservation into sustaining people's livelihoods.

BirdLife has compiled *Threatened Birds of Asia: The BirdLife International Red Data Book* (BirdLife International, 2001), which provides a comprehensive assessment of the status, distribution and conservation needs of threatened birds in Asia, and *Important Bird Areas in Asia: key sites for conservation* (Birdlife International, 2004), which identifies the most important sites in Asia for bird conservation. The Ornithological Society of Pakistan is the BirdLife Affiliate in Pakistan.

Bombay Natural History Society (BNHS)
Hornbill House, Dr Salim Ali Chowk, Shaheed Bhagat Singh Road, Mumbai 400 023, India.
Publications: *Journal of the Bombay Natural History Society, Hornbill* magazine (quarterly).
The BNHS was founded in 1883 and is the largest non-governmental organisation in the subcontinent engaged in the conservation of nature and natural resources, education and research in natural history. The society has an invaluable collection of over 26,000 bird specimens held at its headquarters in Mumbai. The society's Nature Education Wing reaches over 10,000 students each year. The Salim Ali Nature Conservation Fund creates awareness, with training programmes for Indian Army officers, journalists and

trekkers. Scientists have carried out vital ornithological research, including a national bird-ringing programme and studies on grassland birds. BNHS is the BirdLife Partner Designate in India.

Oriental Bird Club (OBC)
P.O. Box 324, Bedford, MK42 0WG, UK.
email: mail@orientalbirdclub.org
Website: www.orientalbirdclub.org
Publications: *BirdingASIA* (half-yearly), *Forktail* journal (annual).
The OBC was established in 1985. It aims to:

- encourage an interest in the birds of the Oriental region and their conservation;
- liaise with and promote the work of existing regional societies;
- collate and publish material on Oriental birds.

The OBC offers grants to nationals in the region to encourage survey work useful for conservation and conservation awareness.

Wetlands International
3A39, Block A, Kelana Centre Point, Jalan SS7/19, Petaling Jaya, Malaysia.
Website: www.wetlands.org
Publications: *Wetlands* (Newsletter of Wetlands International).
Wetlands International is committed to promoting the protection and sustainable utilisation of wetlands and wetland resources worldwide. It has been involved in surveys, monitoring and conservation of waterbirds and wetlands since 1954 and has coordinated the Asian Waterbird Census since 1987. It coordinates the development of flyway initiatives for migratory waterbirds and their habitats including the development of a Central Asian Flyway Action Plan.

World Pheasant Association (WPA)
Website: www.pheasant.org.uk
WPA Pakistan was the first chapter of the World Pheasant Association, having been established in the mid-1970s. Since then, WPA has been involved in a variety of survey and training projects, mainly in the hills rather than the plains. Efforts are currently concentrated in the North West Frontier Province, especially the Palas Valley, and in seeking to provide training in techniques for monitoring pheasants, other species and habitats. Globally, WPA is concerned with stimulating and undertaking field projects to determine the status of species and their conservation needs, determining their importance to rural communities, and developing and implementing conservation action where necessary. It works with BirdLife International and the Galliformes Specialist Groups that it manages with IUCN-SSC to provide the IUCN Red List assessments.

REFERENCES

Alström, P., and Olsson, U. (1999) The Golden-spectacled Warbler – a complex of sibling species, including a previously undescribed species. *Ibis* 141: 545–568.

BirdLife International (2001) *Threatened Birds of Asia*. BirdLife International, Cambridge.

BirdLife International (2004) *Important Bird Areas in Asia: key sites for conservation*. BirdLife International, Cambridge (BirdLife Conservation Series No. 13).

Grimmett, R., Inskipp, C., and Inskipp, T. (1998) *Birds of the Indian Subcontinent*. Christopher Helm, London.

Inskipp, T., Lindsey, N., and Duckworth, W. (1996) *An Annotated Checklist of the Birds of the Oriental Region*. Oriental Bird Club, Sandy.

Klein, R., and Buchheim, A. (1997) Die westliche Schwarzmeerküste als Kontaktgebeit zweier Großmöwenformen der Larus cachinnans-Gruppe. *Vogelwelt* 118: 61–70.

Liebers, D., Helbig, A.J., and de Knijff, P. (2001) Genetic differentiation and phylogeography of gulls in the *Larus cachinnans-fuscus* group (Aves: Charadriiformes). *Molecular Ecology* 10: 2447–2462.

Parry, S.J., Clark, W.S., and Prakash, V. (2002) On the taxonomic status of the Indian Spotted Eagle *Aquila hastata*. *Ibis* 144: 665–675.

Rasmussen, P.C., and Anderton, J.C. (2005) *Birds of South Asia. The Ripley Guide*. 2 vols. Smithsonian Institution and Lynx Edicions, Washington D.C. and Barcelona.

Rasmussen, P.C., and Parry, S.J. (2001) The taxonomic status of the 'Long-billed' Vulture *Gyps indicus*. *Vulture News* 44: 18–21.

Roberts, T.J. (1991) *The Birds of Pakistan, Volume 1. Non-passeriformes*. Oxford University Press, Karachi.

Roberts, T.J. (1991–1992) *The Birds of Pakistan, Volume 2. Passeriformes*. Oxford University Press, Karachi.

Roberts, T.J. (1997) *The Mammals of Pakistan*. Revised edition. Oxford University Press, Karachi.

Stattersfield, A.J., Crosby, M.J., Long, A.J., and Wege, D.C. (1998) *Endemic Bird Areas of the World, Priorities for Biodiversity Conservation*. BirdLife International, Cambridge.

ACKNOWLEDGEMENTS

The authors would like to thank once again those who contributed in a major way to the *Birds of the Indian Subcontinent* (Grimmett *et al.*, 1998) and who are acknowledged in that work. Particular thanks go to Carol Inskipp, the second author of this major work, who due to pressures of other work decided not to co-author the Pakistan guide – this guide could not have been produced without her earlier contribution. Carol nevertheless provided assistance in various ways, including liaison between the authors and with WWF Pakistan with regard to the maps.

The authors are extremely grateful to WWF Pakistan, and in particular Hassan Ali, Salman Ashraf and Uzma Khan, for preparing the maps. Thanks are also due to Oxford University Press for allowing the use of maps, included in the introduction, from T.J. Roberts' guides to the birds and mammals of Pakistan (Roberts 1991, 1997).

We are also extremely grateful to Aleem Ahmed Khan and Imran Khaliq, of the Ornithological Society of Pakistan, for arranging the Urdu edition of this guide, and to M. Zafar-ul Islam, who has greatly assisted with this task by allowing use of his Urdu translation of the *Birds of Northern India*. Thanks also go to Clive Byers, Daniel Cole,

John Cox, Gerald Driessens, Carl D'Silva, Martin Elliott, Kim Franklin, Alan Harris, Peter Hayman, Craig Robson, Jan Wilczur and Tim Worfolk, who prepared the illustrations that are used in this work. Richard Grimmett would also like to thank Helen Taylor, and their two children George and Ella, as well as his parents, Frank and Molly Grimmett, for their continued encouragement and support.

Once again we would also like to thank Nigel Redman of A&C Black, who has supported this project throughout and has overseen the production process, and David and Namrita Price-Goodfellow of D & N Publishing for their expert editing, design and layout of the guide. Ian Woodward and Steve Munns provided assistance with design and mapping.

Finally we would like to thank Kathy MacKinnon and Tony Whitten of the World Bank. They have been champions of national- and local-language field guides for over a decade, in recognition of the vital role they play in helping to promote nature conservation and environmental awareness. The English and Urdu versions of this guide would not have been possible without the support of Kathy, Tony and the World Bank.

GLOSSARY

Acacia scrub forest: dominated by *Acacia modesta*, known locally as Phulai.
Altitudinal migrant: a species that breeds at high altitudes (in mountains) and moves to lower levels and valleys in the non-breeding season.
Arboreal: tree-dwelling.
Arm: the basal part of the wing, from where it joins the body, outwards to the carpal joint.
Artemisia: wormwood, *Artemisia brevifolia* (Syn: *maritima*).
Axillaries: the feathers in the armpit at the base of the underwing.
Ban: Silver Oak, *Quercus incana*.
Banyan: sacred fig tree of India, *Ficus bengalensis*.
Biotope: a particular area that is substantially uniform in its environmental conditions and flora and fauna.
Blue Pine: *Pinus wallichiana*.
Cane Grass: *Saccharum spontaneum* (Syn: *munja*).
Cap: a well-defined patch of colour or bare skin on the top of the head.
Carpal: the bend of the wing or carpal joint.
Carpal patch: a well-defined patch of colour on the underwing in the vicinity of the carpal joint.
Casque: an enlargement on the upper surface of the bill, in front of the head, as in hornbills.
Cere: a fleshy (often brightly coloured) structure at the base of the bill and containing the nostrils.
Chilghoza: edible seed Pine, *Pinus gerardiana*.
Collar: a well-defined band of colour that encircles, or partly encircles, the neck.
Colonial: nesting or roosting in tight colonies; species that are loosely colonial have nests that are more widely spaced.
Commensalism: a rare situation where species A benefits from the presence of species B, but where B is indifferent to the presence of A, neither gaining nor losing from that association.
Culmen: the ridge of the upper mandible.
Deodar: Indian cedar, *Cedrus deodara*.
Dhand: commonly used word for a lake in Sind.
Dwarf Elder: *Sambucus wightiana*.
Eclipse plumage: a female-like plumage acquired in some species (e.g. ducks or some sunbirds) during or after breeding.
Edgings or edges: outer feather margins that can frequently result in distinct paler or darker panels of colour in the wings or tail.
Endemic: restricted or confined to a specific country or region.
False Caper: *Capparis decidua*.

GLOSSARY

Flight feathers: the primaries, secondaries and tail feathers.
Fringes: complete feather margins that frequently result in a scaly appearance to body feathers or coverts.
Frugivorous: fruit-eating.
Gape: the mouth and fleshy corner of the bill, which can extend back below the eye.
Gonys: a bulge in the lower mandible, usually distinct in gulls and terns.
Graduated tail: where the longest tail feathers are the central pair and the shortest are the outermost, with those in between intermediate in length.
Granivorous: feeding on grain or seeds
Gregarious: living in flocks or communities.
Gular pouch: a loose and pronounced area of skin extending from the throat (e.g. in pelicans or hornbills).
Hackles: long and pointed neck feathers that can extend across the mantle and wing-coverts (e.g. in Red Junglefowl).
Hand: the outer end of the wing, from the carpal joint to the tip of the wing.
Headpond: the reservoir created upstream of a barrage, sometimes also partly enclosed by embankments.
Hepatic: used with reference to the rufous-brown morph of some cuckoos.
Himalayan Bird Cherry: *Prunus cornuta*.
Himalayan Peony: *Paeonia emodi*.
Holly Oak: Balut in vernacular, *Quercus ilex*.
Iris (plural irides): the coloured membrane that surrounds the pupil of the eye, which can be brightly coloured.
Irrigation barrage: a concrete or masonry structure built across a river, to hold a series of lock gates to control water flow downstream and enable the feeding of major irrigation canals.
Jheel: a shallow lake in a low-lying natural depression, usually with floating and submerged vegetation, reed-beds and partially submerged trees.
Juniper forest: (tall trees) *Juniperus excelsa*.
Juniper scrub: (Creeping Juniper) *Juniperus communis*.
Kohistan: a place or country of mountains.
Lappet: a wattle, particularly one at the gape.
Leading edge: the edge of the forewing.
Local: occurring or common within a small or restricted area.
Loranthus longiflorus: parasitic creeper.
Mandible: the lower or upper half of the bill.
Mask: a dark area of plumage surrounding the eye and often covering the ear-coverts.
Mazri Palm: *Nannorhops ritchiana*.
Melanistic: when the plumage is dominated by black pigmentation.
Morph: a distinct plumage type, which occurs alongside one or more other distinct plumage types.
Nomadic: of a wandering or erratically occurring species that has no fixed territory when not breeding.
Nominate: the first-named race of a species, whose racial name is the same as the specific name.
Nuchal: the hind-neck, used with reference to a patch or collar.
Nullah: a watercourse or ravine, usually dry.
Ocelli: eye-like spots of iridescent colour; a distinctive feature in the plumage of peafowl.
Orbital ring: a narrow circular ring of feathering or bare skin surrounding the eye.
Pelagic: of the open sea.
pH: a measure of acidity; low pH indicates high acidity, and high pH low acidity.
Phragmites: *Phragmites karka*.
Pipal: sacred fig of India, *Ficus religiosa*.
Pistachio: in Baluchistan, *Pistacia cabulica* and *P. khinjuk*; in Punjab, *Pistacia integerrima*.
Plantation: group of trees (usually exotic or non-native species) planted in close proximity to each other for timber or as a crop.
Primary projection: the extension of the primaries beyond the longest tertial on a closed wing; this can be of critical use (e.g. in the identification of larks or *Acrocephalus* warblers).

Race: subspecies; a geographical population whose members all show constant differences (e.g. in plumage or size) from those of other populations of the same species.
Rectrices: the tail feathers.
Remiges: the primaries and secondaries.
Rictal bristles: bristles, often prominent, at the base of the bill.
Sea Buckthorn: *Hippophae rhamnoides*.
Seepage zone: because of their heavy silt load, the major rivers in Pakistan often have beds that lie above the surrounding lower land. Similarly, unlined major canals built above the surrounding land allow some seepage through their banks. These provide important wet areas in an otherwise desert climate.
Shaft streak: a fine pale or dark line of colour that follows the feather shaft.
Silk Cotton Tree: *Salmalia malabarica* (Syn: *Bombax ceiba*).
Silver Fir: *Abies pindrow*.
Speculum: refers to the often glossy panel across the secondaries of dabbling ducks, which is often bordered by pale tips to these feathers and by a greater-covert wing-bar.
Spruce: *Picea smitheana* (Syn: *morina*).
Subspecies: See Race.
Subterminal band: a dark or pale band, usually broad, that lies behind the terminal band of the tail.
Subtropical Pine: Chir in vernacular, *Pinus roxburghii*.
Terminal band: a dark or pale band, usually broad, at the end of the tail.
Terrestrial: living or occurring mainly on the ground.
Trailing edge: a darker or paler rear edge of the wing.
Vent: the area around the cloaca (anal opening), just behind the legs (should not be confused with the undertail-coverts).
Vermiculated: finely barred or marked with fine or narrow wavy lines, usually visible only at close range.
Viburnum: Ill-scented Viburnum, *Viburnum grandiflorum* (Syn: *V. nervosum*).
Wattle: a lobe of bare, often brightly coloured, skin attached to the head (frequently at the bill base), as in the mynas or wattled lapwings.
Wild Olive: *Olea ferruginea* (Syn: *cuspidata*).
Wing linings: underwing-coverts.
Wing-bar: generally a narrow and well-defined dark or pale bar across the wing; often refers to a band formed by pale tips to the greater or median coverts (or both, as in 'double wing-bar').
Wing-panel: a broader, generally more diffuse pale or dark band across the wing than a wing-bar (often formed by pale edges to the remiges or coverts).

Selected Bibliography

The following are considered to be the most important references for information on the occurrence and status of birds in Pakistan.

Ahmad, N. (1965) The birds of Lahore. *Pakistan J. Sci.* 17: 143–167.
Ali, Sálim (1941) The birds of Bahawalpur (Punjab). *J. Bombay Nat. Hist. Soc.* 42: 704–747.
Ali, Sálim, and Ripley, S.D. (1987) *Compact Handbook of the Birds of India and Pakistan*. Oxford University Press, New Delhi.
Amstutz, B., and Amstutz, M. (1977) *A Checklist of the Birds of Islamabad and Nathia Gali*. Asian Culture Study Group, Islamabad.
Biddulph, J. (1881) The birds of Gilgit. *Stray Feathers* 9: 301–366; *Ibis* (4)5: 35–102.
Biddulph, J. (1882) Further notes on the birds of Gilgit. *Stray Feathers* 10: 157–178; *Ibis* (4)6: 266–290.

BirdLife International (2001) *Threatened Birds of Asia: the Birdlife International Red Data Book.* BirdLife International, Cambridge.

BirdLife International (2003) *Saving Asia's Threatened Birds: a guide for government and civil society.* BirdLife International, Cambridge.

Briggs, F.S., and Osmaston, B.B. (1928) A note on the birds of Peshawar district. *J. Bombay Nat. Hist. Soc.* 32: 744–761.

Butler, E.A. (1877) Additional notes on the birds of Sindh. *Stray Feathers* 5: 322–328.

Butler, E.A. (1878) My last notes on the avifauna of Sind. *Stray Feathers* 7: 173–191.

Butler, E.A. (1879) Further additions to the Sindh avifauna. *Stray Feathers* 8: 386–389.

Christison, A.F.P. (1939) *Handbook of the Birds of Northern Baluchistan.* DHQ Press, Quetta.

Christison, A.F.P. (1941) Notes on the birds of Chagai. *Ibis* (14)5: 531–556.

Christison, A.F.P. (1942) Some additional notes on the distribution of the avifauna of northern Baluchistan. *J. Bombay Nat. Hist. Soc.* 43: 478–487.

Corfield, D.M. (1983) *Birds of Islamabad, Pakistan and the Murree Hills.* Asian Study Group, Islamabad.

Currie, A.J. (1916) The birds of Lahore and the vicinity. *J. Bombay Nat. Hist. Soc.* 24: 561–577.

Delme Radcliffe, H. (1915) List of the birds of Baluchistan. *J. Bombay Nat. Hist. Soc.* 23: 745–757; 24: 156–159.

Eates, K.R. (1968) List of the birds of former Sind and Khairpur (revised in 1934). In: Sorley, H.T. (ed.) *Gazetteer of West Pakistan, the Former Province of Sind.* Government of Pakistan, pp. 108–127.

Fulton, H.T. (1904) Notes on the birds of Chitral. *J. Bombay Nat. Hist. Soc.* 16: 44–64.

Fulton, H.T. (1905) Additional notes on the birds of Chitral. *J. Bombay Nat. Hist. Soc.* 16: 744.

Gilbert, M., Watson, R.T., Virani, M.Z., Oaks, J.L., Ahmed, S., Chaudhry, M.J.I., Arshad, M., Mahmood, S., Ali, A., and Khan, A.A. (2006) Rapid population declines and mortality clusters in three Oriental White-backed Vulture *Gyps bengalensis* colonies in Pakistan due to diclofenac poisoning. *Oryx*: 40(4): 388–399.

Grimmett, R., and Robson, C. (1984a) Annotated list of bird species recorded in Indus Kohistan, May 5–17, 1984. Unpublished appendix to report submitted to National Council for Wildlife Preservation, Government of Pakistan.

Grimmett, R., and Robson, C. (1984b) Annotated list of bird species recorded in Kaghan valley, April 16th–May 29th, 1984. Unpublished appendix to report submitted to National Council for Wildlife Preservation, Government of Pakistan.

Hasan, S.A. (1964) Birds of Manchar Lake. *Agriculture Pakistan* 15: 259–283.

Holmes, D.A., and Wright, J.O. (1968–1969) The birds of Sind: a review. *J. Bombay Nat. Hist. Soc.* 65: 533–566; 66: 8–30.

Hume, A.O. (1873a) Contributions to the ornithology of India – Sindh No. 1. *Stray Feathers* 1: 44–50.

Hume, A.O. (1873b) Contributions to the ornithology of India – Sindh No. 2. *Stray Feathers* 1: 91–289.

Hume, A.O. (1873c) Ornithology. In: Henderson, G., and Hume, A.O. *Lahore to Yarkand*.

Hume, A.O. (1877) Resumé of recent additions to the Sindh avifauna. *Stray Feathers* 5: 328–330.

Jones, A.E. (1921) Bird notes from the Campbellpur-Attock district, western Punjab. *J. Bombay Nat. Hist. Soc.* 27: 794–802.

Khanum, Z., Ahmed, F.M., and Ahmed, M. (1980) A checklist of birds of Pakistan with illustrated keys to their identification. *Rec. Zool. Surv. Pakistan* 9: 1–138.

Khanum, Z., and Qadri, M.A.H. (1972) Fresh records of some birds in West Pakistan. *Pakistan J. Zool.* 4: 219–221.

Koelz, W. (1940) Notes on the winter birds from the lower Punjab. *Pap. Michigan Acad. Sci. Arts Letters* 25: 323–355.

Magrath, H.A.F. (1908) Notes on the birds of Thandiani. *J. Bombay Nat. Hist. Soc.* 18: 284–299.

Magrath, H.A.F. (1909) Bird notes from Murree and the Galis. *J. Bombay Nat. Hist. Soc.* 19: 142–156.

Meinertzhagen, R. (1920) The birds of Quetta. *Ibis* (11)2: 132–195.

Mirza, Z.B. (1973) Notes on the distribution of some birds in West Pakistan. *Pakistan J. Zool.* 5: 203–205.

Mountfort, G. (1969) *The Vanishing Jungle: the story of the World Wildlife Fund expeditions to Pakistan.* Collins, London.

Mountfort, G., and Poore, D. (1968) Report on the Second World Wildlife Fund Expedition to Pakistan. Unpublished report to WWF, Switzerland, 25 pp.

Murray, J.A. (1878) Further additions to the Sindh avifauna. *Stray Feathers* 7: 108–123.

Murray, J.A. (1884) *The Vertebrate Zoology of Sind.* Richards, London.

Murray, J.A. (1887) The zoology of Beloochistan and southern Afghanistan. *Amer. Mus. Nat. Hist.* (?)1: 49–68, 105–131.

Perreau, G.A. (1910) Notes on the birds of Chitral. *J. Bombay Nat. Hist. Soc.* 19: 901–922.

Raja, N.A., Davidson, P., Bean, N., Drijvers, R., Showler, D.A., and Barker, C. (1999) The birds of Palas, North-West Frontier Province, Pakistan. *Forktail* 15: 77–85.

Rattray, R.H. (1899) Birds collected and observed at Thull during five months in 1898, and notes on their nidification. *J. Bombay Nat. Hist. Soc.* 12: 337–348.

Richmond, C.W. (1895) Catalogue of a collection of birds made by Doctor W.L. Abbott in Kashmir, Baltistan and Ladak, with notes on some of the species, and a description of a new species of *Cyanecula*. *Proc. U.S. Nat. Mus.* 18: 451–503.

Ripley, S.D. (1982) *A Synopsis of the Birds of India and Pakistan, Together with those of Nepal, Sikkim, Bhutan and Ceylon.* 2nd edn. Bombay Natural History Society, Bombay.

Roberts, T.J. (1974) Interesting distributional records for Pakistan. *J. Bombay Nat. Hist. Soc.* 70: 552–554.

Roberts, T.J. (1975) Ornithological records for Pakistan. *J. Bombay Nat. Hist. Soc.* 72: 201–204.

Roberts, T.J. (1978) Unusual ornithological records for Pakistan. *J. Bombay Nat. Hist. Soc.* 75: 216–219.

Roberts, T.J. (1980) Bird notes from Baluchistan province, Pakistan. *J. Bombay Nat. Hist. Soc.* 77: 12–20.

Roberts, T.J. (1981) Ornithological notes from Pakistan. *J. Bombay Nat. Hist. Soc.* 78: 73–76.

Roberts, T.J. (1984) Recent ornithological records from Pakistan. *J. Bombay Nat. Hist. Soc.* 81: 399–405.

Roberts, T.J. (1991–1992) *The Birds of Pakistan.* 2 vols. Oxford University Press, Karachi.

Roberts, T.J., Passburg, R., and van Zalinge, N.P. (1980) *A Check List, Birds of Karachi and Lower Sind (Pakistan).* WWF Pakistan, Lahore.

Scully, J. (1881) A contribution to the ornithology of Gilgit. *Stray Feathers* 10: 88–146.

Ticehurst, C.B. (1922–1924) The birds of Sind. *Ibis* (11)4: 526–572, 605–662; (11)5: 1–43, 235–275, 438–474, 645–666; (11)6: 110–146, 495–518.

Ticehurst, C.B. (1926–1927) The birds of British Baluchistan. *J. Bombay Nat. Hist. Soc.* 31: 687–711, 862–881; 32: 64–97.

Ticehurst, C.B. (1930) Birds of Quetta. *J. Bombay Nat. Hist. Soc.* 34: 246–247.

Waite, H.W. (1945) Birds on the Hindustan–Tibet road, N.-W. Himalaya. *J. Bombay Nat. Hist. Soc.* 45: 531–542.

Waite, H.W. (1948) The birds of the Punjab Salt Range (Pakistan). *J. Bombay Nat. Hist. Soc.* 48: 93–117.

Waite, H.W. (1962) Notes on the range of certain birds as given in S.D. Ripley II (1961): 'A synopsis of the birds of India and Pakistan'. *J. Bombay Nat. Hist. Soc.* 59: 959–963.

Whistler, H. (1922) Birds of the Jhang district. *Ibis* (11)4: 259–309, 401–437.

Whitehead, C.H.T. (1910–1911) On the birds of Kohat and the Kurram valley, northern India. *J. Bombay Nat. Hist. Soc.* 20: 169–197, 776–799, 954–980.

Whitehead, C.H.T. (1914) Some notes on the birds of the Kaghan valley, Hazara, North West Frontier Province. *J. Bombay Nat. Hist. Soc.* 23: 104–109.

Williams, C.H., and Williams, C.E. (1929) Some notes on the birds breeding round Quetta. *J. Bombay Nat. Hist. Soc.* 33: 598–613.

Family Summaries

Some families are divided into subfamilies and some of these are further divided into tribes.

■ ■ **Partridges, Francolins, Snowcocks, Quails and Pheasants** Phasianidae ■ ■
Terrestrial, feeding and nesting on the ground, but many species roost in trees at night. They are good runners, often preferring to escape on foot rather than taking to the air. Their flight is powerful and fast, but, except in the case of the migratory quails, cannot be sustained for long periods. Typically, they forage by scratching the ground with their strong feet to expose food hidden among dead leaves or in the soil. They mainly eat seeds, fruit, buds, roots and leaves, complemented by invertebrates. **pp.48–53**

■ ■ **Buttonquails** Turnicidae ■ ■
Small, plump terrestrial birds. They are found in a wide variety of habitats with a dry, often sandy substrate and low ground cover under which they can readily run or walk. Buttonquails are very secretive and fly with great reluctance, with weak whirring beats low over the ground, dropping quickly into cover. They feed on grass and weed seeds, grain, greenery and small insects, picking food from the ground surface, or scratching with the feet. **pp.48–49**

■ ■ **Whistling-ducks** Dendrocygnidae **and Swans, Geese and Ducks** Anatidae ■ ■
Aquatic and highly gregarious, typically migrating, feeding, roosting and resting together, often in mixed flocks. Most species are chiefly vegetarian when adult, feeding on seeds, algae, plants and roots, often supplemented by aquatic invertebrates. Their main foraging methods are diving, surface-feeding or dabbling, and grazing. They also up-end, wade, filter and sieve water and debris for food, and probe with the bill. They have a direct flight with sustained fast wingbeats, and characteristically fly in V-formation. **pp.54–61**

■ ■ **Honeyguides** Indicatoridae ■ ■
Small, inconspicuous birds that inhabit forest or forest edge. All species eat insects, but a peculiarity shared by the family is that they also eat wax, usually as bee combs. They spend long periods perched upright and motionless, and feed by clinging to bee combs, often upside-down, and by aerial sallies. **p.235**

■ ■ **Wrynecks, Piculets and Woodpeckers** Picidae ■ ■
Chiefly arboreal, and usually seen clinging to, and climbing up, vertical trunks and lateral branches. Typically, they work up trunks and along branches in jerky spurts, directly or in spirals. Some species feed regularly on the ground, searching mainly for termites and ants. The bill of many species is powerful, for boring into wood to extract insects and for excavating nest holes. Woodpeckers feed chiefly on ants, termites, and grubs and pupae of wood-boring beetles. Most woodpeckers also hammer rapidly against tree trunks with their bill, producing a loud rattle, known as 'drumming', which is used to advertise and in defence of their territories. Their flight is strong and direct, with marked undulations. Many species can be located by their characteristic loud calls. **pp.62–65**

■ ■ **Asian Barbets** Megalaimidae ■ ■
Arboreal, and usually found in the treetops. Despite their bright coloration, they can be very difficult to see, especially when silent, their plumage blending remarkably well with tree foliage. They often sit motionless for long periods. Barbets call persistently and monotonously in the breeding season, sometimes throughout the day; in the non-breeding season they are usually silent. They are chiefly frugivorous, many species favouring figs (*Ficus* spp.). Their flight is strong and direct, with deep woodpecker-like undulations. **pp.66–67**

■ ■ **Hornbills** Bucerotidae ■ ■
Medium-sized to large birds with massive bills with variable-sized casque. Mainly arboreal, feeding chiefly on wild figs (*Ficus* spp.), berries and drupes, supplemented by small animals and insects. Their flight is powerful and slow, and for most species consists of a few wingbeats followed by a sailing glide with the wing-tips upturned. In all but the smaller species, the wingbeats make a distinctive loud puffing sound that is audible for some distance. Hornbills often fly one after another in follow-my-leader fashion. They are usually found in pairs or small parties, sometimes in flocks of up to 30 or more where food is abundant. **pp.66–67**

■ ■ Hoopoes Upupidae ■ ■

Hoopoes have a distinctive appearance, with a long decurved bill, short legs and rounded wings. They are insectivorous and forage by pecking and probing the ground. Their flight is undulating, slow and butterfly-like. **pp.66–67**

■ ■ Rollers Coraciidae ■ ■

Stoutly built, medium-sized birds with a large head and short neck. Their diet consists mainly of large insects. They usually occur singly or in widely spaced pairs. Their flight is buoyant, with rather rapid deliberate wingbeats. **pp.66–67**

■ ■ Small Kingfishers Alcedinidae, Large Kingfishers Halcyonidae and Pied Kingfishers Cerylidae ■ ■

Small to medium-sized birds, with a large head, long, strong beak and short legs. Most kingfishers spend long periods perched singly or in well-separated pairs, watching intently before plunging swiftly downwards to seize prey with their bill; they usually return to the same perch. They eat mainly fish, tadpoles and invertebrates; larger species also eat frogs, snakes, crabs, lizards and rodents. Their flight is direct and strong, with rapid wingbeats, and often close to the surface. **pp.68–69**

■ ■ Bee-eaters Meropidae ■ ■

Brightly coloured birds with a long decurved beak, pointed wings and very short legs. They catch large flying insects on the wing, by making short, swift sallies like a flycatcher from an exposed perch such as a treetop, branch, post or telegraph wire; insects are pursued in a lively chase with a swift and agile flight. Some species also hawk insects in flight like swallows. Most species are sociable. Their flight is graceful and undulating, and consists of a few rapid wingbeats followed by a glide. **pp.68–69**

■ ■ Old World Cuckoos Cuculidae ■ ■

Old World Cuckoos have an elongated body with a fairly long neck, a tail that varies from medium length to long and graduated, and a quite long, decurved bill. Almost all Cuculidae are arboreal and eat hairy caterpillars. Male cuckoos are very noisy in the breeding season, calling frequently during the day, especially if cloudy, and often into the night. When not breeding they are silent and unobtrusive, and as a result their status and distribution at this season are very poorly known. Cuckoos are notorious for their nest parasitism. **pp.70–73**

■ ■ Coucals Centropodidae ■ ■

Large, skulking birds with a long, graduated tail and weak flight. Coucals are terrestrial, frequenting dense undergrowth, bamboo, tall grassland or scrub jungle. They eat small animals and invertebrates. **pp.72–73**

■ ■ Parrots Psittacidae ■ ■

Parrots have a short neck and short, stout, hooked bill whose upper mandible is strongly curved and overlaps the lower mandible. Most parrots are noisy and highly gregarious. They associate in family parties and small flocks, and gather in large numbers at concentrations of food, such as paddy-fields. Their diet is almost entirely vegetarian, and consists of fruit, seeds, buds, nectar and pollen. The flight of *Psittacula* parrots is swift, powerful and direct. **pp.72–73**

■ ■ Swifts Apodidae ■ ■

Swifts have long, pointed wings, a compact body, a short bill with a wide gape, and very short legs. Swifts spend most of the day swooping and wheeling in the sky with great agility and grace. Typical swift flight is a series of rapid shallow wingbeats interspersed with short glides. They feed entirely in the air, drink and bathe while swooping low over water, and regularly pass the night in the air. Swifts eat mainly tiny insects, caught by flying back and forth among aerial concentrations of these with their large mouth open; they also pursue individual insects. **pp.74–75**

FAMILY SUMMARIES

■ ■ Barn Owls Tytonidae and Typical Owls Strigidae ■ ■
Owls have a large and rounded head, big forward-facing eyes surrounded by a broad facial disc, and a short tail. Most are nocturnal and cryptically coloured and patterned, making them inconspicuous when resting during the day. When hunting, owls either quarter the ground or scan and listen for prey from a perch. Their diet consists of small animals and invertebrates. Owls are usually located by their distinctive and often weird calls, which are diagnostic of the species and advertise their presence and territories. **pp.76–79**

■ ■ Nightjars Caprimulgidae ■ ■
Small to medium-sized birds with long, pointed wings, and a gaping mouth with long bristles that help to catch insects in flight. Nightjars are crepuscular and nocturnal in habit, with soft, owl-like, cryptically patterned plumage. By day they perch on the ground or lengthwise on a branch, and are difficult to detect. They eat flying insects that are caught on the wing. Typically, they fly erratically to and fro over and among vegetation, occasionally wheeling, gliding and hovering to pick insects from foliage. They are most easily located by their calls. **pp.80–81**

■ ■ Pigeons and Doves Columbidae ■ ■
Birds with a stout, compact body, rather short neck, and small head and bill. Their flight is swift and direct, with fast wingbeats. Most species are gregarious outside the breeding season. Seeds, fruits, buds and leaves form their main diet, but many species also eat small invertebrates. They have soft plaintive cooing or booming voices, and the calls are often monotonously repeated. **pp.82–85**

■ ■ Bustards Otididae ■ ■
Medium-sized to large terrestrial birds that inhabit grasslands, semi-desert and desert. They have fairly long legs, a stout body, long neck, and crests and neck plumes, which are exhibited in display. The wings are broad and long, and in flight the neck is outstretched. Their flight is powerful and can be very fast. When feeding, bustards have a steady, deliberate gait. They are more or less omnivorous, and feed opportunistically on large insects, such as grasshoppers and locusts, young birds, shoots, leaves, seeds and fruits. Males perform elaborate and spectacular displays in the breeding season. **pp.86–87**

■ ■ Cranes Gruidae ■ ■
Stately long-necked, long-legged birds with a tapering body, and long inner secondaries that hang over the tail. The flight is powerful, with the head and neck extended forwards and legs and feet stretched out behind. Flocks of cranes often fly in V-formation; they sometimes soar at considerable heights. Most cranes are gregarious outside the breeding season, and flocks are often very noisy. Cranes have a characteristic resonant and far-reaching musical trumpet-like call. A wide variety of plant and animal food is taken. The bill is used to probe and dig for plant roots, and to graze and glean vegetable material above the ground. Both sexes have a spectacular and beautiful dance that takes place throughout the year. **pp.88–89**

■ ■ Rails, Gallinules and Coots Rallidae ■ ■
Small to medium-sized birds, with moderate to long legs for wading, and short, rounded wings. With the exception of the Common Moorhen and Common Coot, which spend much time swimming in the open, rails are mainly terrestrial. Many occur in marshes. They fly reluctantly and feebly, with legs dangling, for a short distance and then drop into cover again. The majority are heard more often than seen, and are most voluble at dusk and at night. Their calls consist of strident or raucous repeated notes. They eat insects, crustaceans, amphibians, fish and vegetable matter. **pp.88–91**

■ ■ Sandgrouse Pteroclidae ■ ■

Cryptically patterned terrestrial birds resembling pigeons in size and shape. The wings are long and pointed. Most sandgrouse are wary and, when disturbed, rise with a clatter of wings, flying off rapidly and directly with fast, regular wingbeats. They walk and run well, foraging mainly for small, hard seeds picked up from the ground and sometimes also eating green leaves, shoots, berries, small bulbs and insects. They need to drink every day, and will sometimes travel distances to waterholes. Most sandgrouse have regular drinking times, which are characteristic of each species, and they often visit traditional watering places, sometimes gathering in large numbers. Most species are gregarious except when breeding. **pp.92–93**

■ ■ Woodcocks, Snipes, Godwits, Sandpipers, Curlews and Phalaropes Scolopacidae ■ ■

Woodcocks and Snipes Subfamily Scolopacinae

Woodcocks and snipes are small to medium-sized waders with a very long bill, fairly long legs and cryptically patterned plumage. They feed mainly by probing in soft ground and also by picking from the surface. Their diet consists mostly of small aquatic invertebrates. If approached, they usually first crouch on the ground and 'freeze', preferring to rely on their protective plumage pattern to escape detection. They inhabit marshy ground.
pp.94–95

Godwits, Sandpipers, Curlews and Phalaropes Subfamily Tringinae

The Tringinae are wading birds with quite long to very long legs and a long bill. They feed on small aquatic invertebrates. **pp.96–101**

■ ■ Painted-snipes Rostratulidae ■ ■

Painted-snipes frequent marshes and superficially resemble snipes, but have spectacular plumages. **pp.94–95**

■ ■ Jacanas Jacanidae ■ ■

Jacanas characteristically have very long toes, which enable them to walk over floating vegetation. They inhabit freshwater lakes, ponds and marshes. **pp.102–103**

■ ■ Thick-knees Burhinidae ■ ■

Medium-sized to large waders, mainly crepuscular or nocturnal, and with cryptically patterned plumages. They eat invertebrates and small animals. **pp.102–103**

■ ■ Oystercatchers, Ibisbill, Avocets, Plovers and Lapwings Charadriidae ■ ■

Oystercatchers, Ibisbill, Avocets and Stilts Subfamily Recurvirostrinae

Oystercatchers are waders that usually inhabit the seashore and are only vagrants inland. They have all-black or black and white plumage. The bill is long, stout, orange-red and adapted for opening shells of bivalve molluscs. Stilts and avocets have a characteristic long bill, and longer legs in proportion to the body than any other birds except flamingos. They inhabit marshes, lakes and pools. The Ibisbill has a distinctive decurved bill and frequents rivers and streams. All these birds feed on aquatic invertebrates. **pp.102–103**

Plovers and Lapwings Subfamily Charadriinae

Plovers and lapwings are small to medium-sized waders with a rounded head, short neck and short bill. Typically, they forage by running in short spurts, pausing and standing erect, then stooping to pick up invertebrate prey. Their flight is swift and direct. **pp.104–107**

■ ■ Crab-plover, Pratincoles and Coursers, Glareolidae ■ ■

Crab-plover Subfamily Dromadinae

The Crab-plover is the only species in this subfamily. It is usually found singly, in pairs and in small parties, but hundreds may occur at traditional roost sites, and is mainly crepuscular. It feeds chiefly on crabs, mudskippers and crustaceans, which it hunts in a plover-like manner. **pp.102–103**

FAMILY SUMMARIES

Pratincoles and Coursers Subfamily Glareolinae

Coursers and pratincoles have an arched and pointed bill, wide gape and long, pointed wings. Coursers are long-legged and resemble plovers; they feed on the ground. The majority of pratincoles are short-legged; they catch most of their prey in the air, although they also feed on the ground. All pratincoles live near water, whereas coursers frequent dry grassland and dry stony areas. **pp.108–109**

■ ■ Jaegers, Skimmers, Gulls and Terns Laridae ■ ■

This family comprises the subfamilies Larinae and Alcinae. Only the subfamily Larinae occurs in Pakistan, with representatives from four tribes occurring in the region.

Jaegers Tribe Stercorariini

Aerial seabirds, with a strong, hooked beak, long, pointed wings, short legs and webbed feet. Jaegers feed by chasing other seabirds, especially terns, until they drop or disgorge their food. They are usually found in marine waters, some distance from land, but are occasionally found inshore and may occur inland after monsoon storms. **pp.110–111**

Skimmers Tribe Rynchopini

Skimmers have very long wings, a short, forked tail, a long bill and short red legs and toes, and are black above and white below. They frequent rivers and lakes. **pp.108–109**

Gulls Tribe Larini

Medium-sized to large birds with relatively long, narrow wings, usually a stout bill, moderately long legs and webbed feet. Immatures are brownish and cryptically patterned. In flight, gulls are graceful and soar easily in updraughts. Most species are gregarious, and all species swim buoyantly and well. They are highly adaptable, and most species are opportunistic feeders with a varied diet, including invertebrates. **pp.110–113**

Terns Tribe Sternini

Small to medium-sized aerial birds with a gull-like body, but generally more delicately built. The wings are long and pointed, typically narrower than those of the gulls, and the flight is buoyant and graceful. Terns are highly vocal and most species are gregarious. Two main groups of terns occur in Pakistan: the *Sterna* terns and the *Chlidonias* or marsh terns. The *Sterna* terns have a deeply forked tail. *Sterna* terns mainly eat small fish, tadpoles and crabs caught by hovering and then plunge-diving from the air, often submerging completely, and also by picking prey from the surface. Marsh terns lack a prominent tail-fork and, compared with *Sterna* terns, are smaller, more compact and short-tailed, and have a more erratic and rather stiff-winged flight. Typically, marsh terns hawk insects or swoop down to pick small prey from the water surface. **pp.112–117**

■ ■ Osprey, Hawks, Eagles, Harriers and Vultures etc. Accipitridae ■ ■

A large and varied family of raptors, ranging from the Eurasian Sparrowhawk to the huge Himalayan Griffon. In most species, the vultures being an exception, the female is larger than the male and is often duller and brownish. The Accipitridae feed on mammals, birds, reptiles, amphibians, fish, crabs, molluscs and insects – dead or alive. All have a hooked, pointed bill and very acute sight, and all except the vultures have powerful feet with long, curved claws. They frequent all habitat types, ranging from dense forest, deserts and mountains to fresh waters. **pp.118–135**

■ ■ Falcons Falconidae ■ ■

Small to medium-sized birds of prey, which resemble the Accipitidrae in having a hooked beak, sharp, curved talons, and remarkable powers of sight and flight. Like other raptors they are mainly diurnal, although a few are crepuscular. Some falcons kill flying birds in a surprise attack, often by swooping at great speed (e.g. Peregrine); others hover and then swoop on prey on the ground (e.g. Common Kestrel), and several species hawk insects in flight (e.g. Eurasian Hobby). **pp.136–139**

■ ■ Grebes Podicipedidae ■ ■

Aquatic birds adapted for diving from the surface and swimming under the water to catch fish and aquatic invertebrates. Their strong legs are placed near the rear of their almost tailless body, and the feet are lobed. In flight, grebes have an elongated appearance, with the neck extended and feet hanging lower than the humped back. They usually feed singly, but may form loose congregations in the non-breeding season.   **pp.140–141**

■ ■ Tropicbirds Phaethontidae ■ ■

Tropicbirds are aerial seabirds, which range over tropical and subtropical waters and nest mainly on oceanic and offshore islands. Their flight is graceful and pigeon-like, with flapping and circling alternating with long glides. They are usually solitary, but may congregate with flocks of feeding terns. They feed by first hovering to locate prey (mainly fish and squid), and then plunge-diving on half-closed wings.   **pp.150–151**

■ ■ Boobies Sulidae ■ ■

Boobies are large seabirds. They forage on the wing, scanning the sea, and on sighting fish or squid they plunge-dive at an angle. Their flight is direct, with alternating periods of flapping and gliding.   **pp.150–151**

■ ■ Anhingas Anhingidae ■ ■

Anhingas are large aquatic birds, adapted for hunting fish underwater. They have a long, slender neck and head, long wings and a very long tail.   **pp.140–141**

■ ■ Cormorants Phalacrocoracidae ■ ■

Cormorants are long-necked, medium-sized to large aquatic birds, with a hooked bill of moderate length and a long, stiff tail. They swim with the body low in the water, with the neck straight, and with the head and bill pointing a little upwards. They eat mainly fish, which are caught by underwater pursuit. In flight, the neck is extended and the head is held slightly above the horizontal. Typically, they often perch for long periods in an upright posture, with spread wings and tail, on trees, posts or rocks.   **pp.140–141**

■ ■ Herons and Bitterns Ardeidae ■ ■

Medium-sized to large birds with long legs for wading. The diurnal herons have a slender body and long head and neck; the night herons are more squat, with a shorter neck and legs. They fly with leisurely flaps, with the legs outstretched and projecting beyond the tail, and nearly always with the head and neck drawn back. They frequent marshes and the shores of lakes and rivers. Typically, herons feed by standing motionless at the water's edge, waiting for prey to swim within reach, or by slow stalking in shallow water or on land. Bitterns usually skulk in reed-beds, although occasionally one may forage in the open, and they can clamber about reed stems with agility. Normally they are solitary and crepuscular and most often seen flying low over reed-beds with slow wingbeats, soon dropping into cover again. When in danger, bitterns freeze, pointing the bill and neck upwards and compressing their feathers so that the whole body appears elongated. Bitterns are characterised by their booming territorial calls. Herons and bitterns feed on a wide variety of aquatic prey.   **pp.142–145**

■ ■ Flamingos Phoenicopteridae ■ ■

Flamingos are large wading birds with a long neck, very long legs, webbed feet and pink plumage. The bill is highly specialised for filter-feeding. Flamingos often occur in huge numbers and are found mainly on salt lakes and lagoons.   **pp.146–147**

■ ■ Ibises and Spoonbills Threskiornithidae ■ ■

These are large birds with a long neck and legs, partly webbed feet and long, broad wings. Ibises have a long decurved bill, and forage by probing in shallow water, mud and grass. Spoonbills have a long spatulate bill, and catch floating prey in shallow water.   **pp.146–147**

■ ■ Storks Ciconiidae ■ ■

Large or very large birds with long bill, neck and legs, long and broad wings and a short tail. In flight, the legs are extended behind and the neck is outstretched. They have a powerful slow-flapping flight and frequently soar for long periods, often at great heights. They capture fish, frogs, snakes, lizards, large insects, crustaceans and molluscs while walking slowly in marshes, at edges of lakes and rivers and in grassland.   **pp.148–149**

■ ■ Pelicans Pelecanidae ■ ■

Pelicans are large, aquatic, gregarious, fish-eating birds. The wings are long and broad, and the tail is short and rounded. They have a characteristic long, straight, flattened bill, hooked at the tip, and with a large expandable pouch suspended beneath the lower mandible. Pelicans often fish cooperatively by swimming forward in a semicircular formation, driving the fish into shallow water; each bird then scoops up fish from the water into its pouch, before swallowing the food. Pelicans fly either in V-formation or in lines, and often soar for considerable periods in thermals. They are powerful fliers, proceeding by steady flaps and with the head drawn back between the shoulders. When swimming, the closed wings are typically held above the back.   **pp.150–151**

■ ■ Shearwaters and Storm-petrels Procellariidae ■ ■

These are marine species, coming to shore only to breed. They are typically gregarious, often gathering in flocks at food concentrations. They feed on zooplankton, squid, fish and offal, seized on or below the water surface. Shearwaters typically fly by a combination of rapid, rather stiff wingbeats, interspersed with long glides (gliding, or 'shearing', being more pronounced in strong winds). The storm-petrels have a more fluttering flight.
   pp.150–151

■ ■ Pittas Pittidae ■ ■

Pittas are brilliantly coloured terrestrial forest passerines. They are of medium size, stocky and long-legged, with a short, square tail, stout bill and an erect carriage. Most of their time is spent foraging for invertebrates on the forest floor, flicking leaves and other vegetation, and probing with their strong bill into leaf litter and damp earth. Pittas usually progress on the ground by long hopping bounds. Typically, they are skulking and are often most easily located by their high-pitched whistling calls or songs. They sing in trees or bushes.   **pp.152–153**

■ ■ Broadbills Eurylaimidae ■ ■

Small to medium-sized plump birds with rounded wings and short legs; most species have a distinctively broad bill. Typically, they inhabit the middle storey of forests and feed mainly on invertebrates gleaned from leaves and branches. Broadbills are active when foraging, but are often unobtrusive and lethargic at other times.   **p.238**

■ ■ Shrikes Laniidae ■ ■

Shrikes are medium-sized, predatory passerines with a strong, stout bill, hooked at the tip of the upper mandible, and have strong legs and feet, a large head, and a long tail with a graduated tip. They search for prey from a vantage point, such as the top of a bush or small tree or post, and swoop down to catch invertebrates or small animals from the ground or in flight. Over long distances their flight is typically undulating. Their calls are harsh, but most have quite musical songs and are good mimics. Shrikes typically inhabit open country with scattered bushes or light scrub.   **pp.152–153**

■ ■ Corvids Corvidae ■ ■

This is very large family, represented in Pakistan by three subfamilies (in some cases further subdivided into tribes).

Subfamily Corvinae

Jays, Magpies, Treepies, Choughs, Nutcrackers, Crows and Ravens Tribe Corvini

These are all robust perching birds, which differ considerably from each other in appearance but have a number of features in common: a fairly long, straight bill, very strong feet and legs, and a tuft of nasal bristles extending over the base of the upper mandible. The sexes are alike or almost alike in plumage. They are strong fliers. Most are gregarious, especially when feeding and roosting. Typically, they are noisy birds, uttering loud and discordant squawks, croaks or screeches. The Corvini are highly inquisitive and adaptable. **pp.154–157**

Orioles, Cuckooshrikes and Minivets Tribe Oriolini

Orioles Genus *Oriolus* These are medium-sized arboreal passerines that usually keep hidden in the leafy canopy. Orioles have beautiful, fluty, whistling songs and harsh, grating calls. They are usually seen singly, in pairs or in family parties. Their flight is powerful and undulating, with fast wingbeats. They feed mainly on insects and fruit. **pp.158–159**

Cuckooshrikes Genus *Coracina* Cuckooshrikes are arboreal, insectivorous birds that usually keep high in the trees. They are of medium size, with long, pointed wings, a moderately long, rounded tail, and an upright carriage when perched. **pp.158–159**

Minivets Genus *Pericrocotus* Small to medium-sized brightly coloured passerines with a moderately long tail and an upright stance when perched. They are arboreal, and feed on insects by flitting about in the foliage to glean prey from leaves, buds and bark, sometimes hovering in front of a sprig or making short aerial sallies. They usually keep in pairs in the breeding season, and in small parties when not breeding. When feeding and in flight, they continually utter contact calls. **pp.158–159**

Subfamily Dicrurinae

Fantails Tribe Rhipidurini

Fantails are small, confiding, arboreal birds, perpetually on the move in search of insects. Characteristically, they erect and spread the tail like a fan, and droop the wings, while pirouetting and turning from side to side with jerky, restless movements. When foraging, they flit from branch to branch, making frequent aerial sallies after winged insects. They call continually. Fantails are usually found singly or in pairs, and often join mixed hunting parties with other insectivorous birds. **pp.158–159**

Drongos Tribe Dicrurini

Drongos are medium-sized passerines with characteristic black and often glossy plumage, a long, often deeply forked tail, and a very upright stance when perched. They are mainly arboreal and insectivorous, catching large winged insects by aerial sallies from a perch. They are usually found singly or in pairs. Their direct flight is swift, strong and undulating. Drongos are rather noisy, and have a varied repertoire of harsh calls and pleasant whistles; some species are good mimics. **pp.160–161**

Monarchs Tribe Monarchini

Most species are small to medium-sized, with long, pointed wings and a medium-length to long tail. They feed mainly on insects. **p.238**

Woodshrikes Subfamily Malaconotinae

Woodshrikes are medium-sized, arboreal, insectivorous passerines. The bill is stout and hooked, the wings are rounded and the tail is short. **pp.160–161**

■ ■ Waxwings Bombycillidae ■ ■

Waxwings have soft plumage, a crested head, a short, broad-based bill, and short, strong legs and feet. Outside the breeding season, they occur in flocks.

p.238

■ ■ Dippers Cinclidae ■ ■

Dippers have short wings and tail, and are adapted for feeding on invertebrates in or under running water. They fly low over the water surface on rapidly whirring wings.

pp.160–161

■ ■ Thrushes, Old World Flycatchers and Chats Muscicapidae ■ ■

This is a large and varied family, represented in Pakistan by two subfamilies, the second of which is subdivided into two tribes.

Subfamily Turdinae

Thrushes Genera *Monticola, Zoothera* and *Turdus*

These are medium-sized passerines with rather long, strong legs, a slender bill and fairly long wings. On the ground they progress by hopping. All are insectivorous, and many eat fruit as well. Some species are chiefly terrestrial and others are arboreal. Most thrushes have loud and varied songs, which are used to proclaim and defend their territories when breeding. Many species gather in flocks outside the breeding season.

pp.162–167

Subfamily Muscicapinae

Old World Flycatchers Tribe Muscicapini

Small insectivorous birds with a small, flattened bill, and bristles at the gape that help in the capture of flying insects. They normally have a very upright stance when perched. Many species frequently flick the tail and hold the wings slightly drooped. Generally, flycatchers frequent trees and bushes. Some species regularly perch on a vantage point, from which they catch insects in mid-air in short aerial sallies or by dropping to the ground, often returning to the same perch. Other species capture insects while flitting among branches or by picking them from foliage. Flycatchers are usually found singly or in pairs; a few join mixed hunting parties of other insectivorous birds.

pp.168–171

Chats Tribe Saxicolini

A diverse group of small to medium-sized passerines that includes the chats, bush robins, magpie robins, redstarts, forktails, cochoas and wheatears. Most are terrestrial or partly terrestrial, some are arboreal, and some are closely associated with water. Their main diet is insects, and they also consume fruits, especially berries. They forage mainly by hopping about on the ground in search of prey, or by perching on a low vantage point and then dropping to the ground onto insects or making short sallies to catch them in the air. They are found singly or in pairs.

pp.172–181

■ ■ Starlings and Mynas Sturnidae ■ ■

Robust, medium-sized passerines with strong legs and bill, moderately long wings and a square tail. The flight is direct; it is strong and fast in the more pointed-winged species (*Sturnus*), and rather slower with more deliberate flapping in the more rounded-winged ones. Most species walk with an upright stance in a characteristic, purposeful jaunty fashion, broken by occasional short runs and hops. Their calls are often loud, harsh and grating, and the song of many species is a variety of whistles; mimicry is common. Most are highly gregarious at times. Some starlings are mainly arboreal and feed on fruits and insects; others are chiefly ground-feeders and are omnivorous. Many are closely associated with human cultivation and habitation.

pp.182–183

■ ■ Nuthatches and Wallcreeper Sittidae ■ ■

Nuthatches and the Wallcreeper are small, energetic passerines with a compact body, short tail, large, strong feet and a long bill. The Wallcreeper is adept at clambering over rock faces. Nuthatches are also agile tree climbers. They can move with ease upwards, downwards, sideways and upside-down over trunks or branches, progressing by a series of jerky hops. Unlike woodpeckers and treecreepers, they usually begin near the top of a tree and work down the main trunk or larger branches, often head-first, and do not use the tail as a prop. Their flight is direct over short distances, and undulating over longer ones. Nuthatches capture insects and spiders, and also eat seeds and nuts. They are often found singly or in pairs; outside the breeding season, they often join foraging flocks of other insectivorous birds.

pp.184–185

■ ■ Treecreepers and Wrens Certhiidae ■ ■

A family of mainly small, rather similar-looking species with two subfamilies.

Treecreepers Subfamily Certhiinae

Treecreepers are small, quiet, arboreal passerines, with a slender, decurved bill and a stiff tail that is used as a prop when climbing, like that of the woodpeckers. They forage by creeping up vertical trunks and along the underside of branches, spiralling upwards in a series of jerks in search of insects and spiders; on reaching the top of a tree, they fly to the base of the next one. Their flight is undulating and weak, and is usually only over short distances. Treecreepers are non-gregarious, but outside the nesting season they usually join mixed hunting parties of other insectivorous birds. They inhabit broadleaved and coniferous forest, woodland, groves, and gardens with trees. Thin, high-pitched contact calls are used continually.

pp.184–185

Wrens Subfamily Troglodytinae

Wrens are small, plump, insectivorous passerines with rather short, blunt wings and strong legs, and with the tail characteristically held erect.

pp.184–185

■ ■ Tits Paridae and Long-tailed Tits Aegithalidae ■ ■

Tits are small, active, highly acrobatic passerines with a short bill and strong feet. Their flight over long distances is undulating. They are mainly insectivorous, although many species also depend on seeds, particularly from trees in winter, and some also eat fruit. They probe bark crevices, search branches and leaves, and frequently hang upside-down from twigs. Tits are chiefly arboreal, but also descend to the ground to feed, hopping about and flicking aside leaves and other debris. They are very gregarious; in the non-breeding season most species join roving flocks of other insectivorous birds.

pp.186–187

■ ■ Swallows and Martins Hirundinidae ■ ■

Hirundines are gregarious, rather small passerines with a distinctive slender, streamlined body, long, pointed wings and a small bill. The long-tailed species are often called swallows and the shorter-tailed species termed martins. All hawk day-flying insects in swift, agile, sustained flight, sometimes high in the air, and catch most of their food while flying in the open. Many species have a deeply forked tail, which affords better manoeuvrability. They perch readily on exposed branches and wires.

pp.188–191

■ ■ Kinglets Regulidae ■ ■

Kinglets are tiny passerines with bright crown feathers, and are represented by only one species in Pakistan: Goldcrest. Typically, they inhabit the canopy of coniferous forest, frequently hovering to catch insects. They are often in mixed feeding parties.

pp.202–203

■ ■ Bulbuls Pycnonotidae ■ ■

Medium-sized passerines with soft, fluffy plumage, rather short and rounded wings, a medium-long to long tail, a slender bill and short, weak legs. Bulbuls feed on berries and other fruits, often supplemented by insects, and sometimes also nectar and buds of trees and shrubs. Many species are noisy, especially when feeding. Typically, bulbuls have a variety of cheerful, loud, chattering, babbling and whistling calls. Most species are gregarious in the non-breeding season.

pp.190–191

FAMILY SUMMARIES

■ ■ Grey Hypocolius Hypocoliidae ■ ■
Grey Hypocolius is the only member of this family. It forages chiefly by hopping and clambering about within trees and bushes, feeding mainly on berries. The flight is strong and direct, with rapid wingbeats and occasional swooping glides or high circling. It settles on the tops of bushes and remains still for long periods, and raises its nape feathers when excited or alarmed. It is often found in flocks in winter. **pp.190–191**

■ ■ Cisticolas, Prinias and Streaked Scrub Warbler Cisticolidae ■ ■

Cisticolas Genus *Cisticola* This genus contains tiny, short-tailed, insectivorous passerines. The tail is longer in winter than in summer. They are often found in grassy habitats, and many have aerial displays. **pp.194–195**

Prinias Genus *Prinia* Prinias have a long, graduated tail, which is longer in winter than in summer. Most inhabit grassland, marsh vegetation or scrub. They forage by gleaning insects and spiders from vegetation, and some species also feed on the ground. When perched, the tail is often held cocked and slightly fanned. The flight is weak and jerky. **pp.192–195**

Streaked Scrub Warbler Genus *Scotocerca* Similar in appearance to the prinias, with a long tail, often cocked. Typically, they forage on the ground for insects in a rather secretive manner. **pp.194–195**

■ ■ White-eyes Zosteropidae ■ ■

White-eyes are small insectivorous passerines with a slightly decurved and pointed bill, a brush-tipped tongue, and a white ring around each eye. White-eyes frequent forest, forest edge, and bushes in gardens. **pp.194–195**

■ ■ Warblers, Laughingthrushes and Babblers Sylviidae ■ ■
A huge and varied family of mostly small species.

Warblers Subfamily *Acrocephalinae*
A large group of small, active perching birds with a fine pointed bill. Insects and spiders form their main diet; some species also consume berries, seeds and nectar. They usually capture their prey by gleaning from foliage, but sometimes also from the ground. Warblers inhabit all types of vegetation, often in dense habitats.

Bush Warblers Medium-sized warblers with rounded wings and tail, and inhabit marshes, grassland and forest undergrowth. They are usually found singly. Bush warblers call frequently, and are usually heard more often than seen. *Cettia* species have surprisingly loud voices, and some can be identified by their distinctive melodious songs. Bush warblers seek insects and spiders by actively flitting and hopping about in vegetation close to the ground. They are reluctant to fly, and usually cover only short distances at low level before dropping into dense cover again. When excited, they flick their wings and tail.
pp.198–199

Locustella Warblers *Locustella* warblers are very skulking, medium-sized warblers with a rounded tail, and are usually found singly. Characteristically, they keep low down or on the ground among dense vegetation, walking furtively and scurrying off when startled. They fly at low level, flitting between plants, or rather jerkily over longer distances, ending in a sudden dive into cover. **pp.198–199**

Acrocephalus Warblers Medium-sized to large warblers, with a prominent bill and rounded tail. They usually occur singly. Many species are skulking, typically keeping low down in dense vegetation. Most frequent marshy habitats, and are able to clamber about readily in reeds and other vertical stems of marsh plants. Their songs are harsh and often monotonous. **pp.198–199**

Hippolais Warblers *Hippolais* warblers are medium-sized warblers with a large bill, a square-ended tail, and a rather sloping forehead and peaked crown that give a distinctive domed head shape. Their songs are harsh and varied. They clamber about vegetation with a rather clumsy action. **pp.200–201**

Tailorbirds Tailorbirds have a long, decurved bill, short wings and a graduated tail, the latter held characteristically cocked. **pp.194–195**

Tit Warblers These are small warblers with soft, copious plumage. They inhabit scrub or coniferous forest. **pp.194–195**

Phylloscopus Warblers *Phylloscopus* warblers are rather small, slim and short-billed warblers. Useful identification features are voice, strength of supercilium, colour of underparts, rump, bill and legs, and presence or absence of wing-bars, of coronal bands or of white on the tail. The coloration of the upperparts and underparts and the presence or prominence of wing-bars are affected by wear. Leaf warblers are fast-moving and restless, hopping and creeping about actively and often flicking their wings. They mostly glean small insects and spiders from foliage, twigs and branches, often first disturbing prey by hovering and fluttering; they also make short fly-catching sallies. **pp.202–203**

Seicercus Warblers *Seicercus* warblers are small and active. They feed in a similar manner to *Phylloscopus* warblers, by gleaning from foliage and twigs and making frequent aerial sallies, but have a broader bill and brighter plumage than those species. **pp.202–203**

Grassbirds Subfamily Megalurinae
Grassbirds are brownish warblers with a longish tail, and inhabit damp tall grassland. The males perform song flights in the breeding season. **pp.194–195**

Laughingthrushes Subfamily Garrulacinae
Laughingthrushes are medium-sized, long-tailed passerines that are gregarious even in the breeding season. At the first sign of danger, they characteristically break into a concert of loud hissing, chattering and squealing. They often feed on the ground, moving along with long, springy hops, rummaging among leaf litter, flicking leaves aside and into the air, and digging for food with their strong bill. Their flight is short and clumsy, with the birds flying from tree to tree in follow-my-leader fashion. **pp.204–205**

Babblers Subfamily Sylviinae, Tribe Timaliini
A large and diverse group of small to medium-sized passerines. They have soft, loose plumage, short or fairly short wings, and strong feet and legs. The sexes are alike in most species. Members of this tribe associate in flocks outside the breeding season, and some species do so throughout the year. Babbler flocks are frequently a component of mixed-species feeding parties. Most babblers have a wide range of chatters, rattles and whistles; some have a melodious song. Many are terrestrial or inhabit bushes or grass close to the ground, while other species are arboreal. Babblers are chiefly insectivorous, and augment their diet with fruits, seeds and nectar. Arboreal species collect food from leaves, moss, lichen and bark; terrestrial species forage by probing, digging and tossing aside dead foliage. **pp.204–207**

Sylvia Warblers Subfamily Sylviinae, Tribe Sylviini
Small to medium-sized passerines with a fine bill, closely resembling the true warblers. Typically, they inhabit bushes and scrub, and feed chiefly by gleaning insects from foliage and twigs; they sometimes also consume berries in autumn and winter. **pp.196–197**

■ ■ Larks Alaudidae ■ ■

Larks are terrestrial, cryptically coloured passerines, generally small in size. They usually walk and run on the ground and often have a very elongated hindclaw. Their flight is strong and undulating. Larks take a wide variety of food, including insects, molluscs, arthropods, seeds, flowers, buds and leaves. Many species have a melodious song, which is often delivered in a distinctive, steeply climbing or circling aerial display, but also from a conspicuous low perch. They live in a wide range of open habitats, including grassland and cultivation. **pp.208–211**

■ ■ Flowerpeckers and Sunbirds Nectariniidae. Subfamily Nectariniinae ■ ■

These birds are represented in Pakistan by two discrete tribes.

Flowerpeckers Tribe Dicaeini

Flowerpeckers are very small passerines with a short beak and tail, and a tongue adapted for nectar-feeding. They usually frequent the tree canopy and feed mainly on soft fruits, berries and nectar, and on small insects and spiders. Many species are especially fond of mistletoe (*Loranthus*) berries. Flowerpeckers are very active, continually flying about restlessly, and twisting and turning in different attitudes when perched, while calling frequently with high-pitched notes. Normally they live singly or in pairs; some species form small parties in the non-breeding season. **pp.212–213**

Sunbirds Tribe Nectariniini

Sunbirds have a bill and tongue adapted to feed on nectar; they also eat small insects and spiders. The bill is long, thin and curved for probing the corollas of flowers. The tongue is very long, tubular and extensible far beyond the bill, and is used to draw out nectar. Sunbirds feed mainly at the blossoms of flowering trees and shrubs. They flit and dart actively from flower to flower, clambering over the blossoms, often hovering momentarily in front of them, and clinging acrobatically to twigs. Sunbirds usually keep singly or in pairs, although several may congregate in flowering trees, and some species join mixed foraging flocks. They have sharp, metallic calls and high-pitched trilling and twittering songs. **pp.212–213**

■ ■ Passeridae ■ ■

Represented in Pakistan by five subfamilies: sparrows, wagtails and pipits, accentors, weavers and estrildine finches.

Sparrows Subfamily Passerinae

Sparrows are small passerines with a thick conical bill. This subfamily includes *Passer*, the true sparrows, some of which are closely associated with human habitation, and *Petronia*, the rock sparrows, which inhabit dry rocky country or light scrub. Most species feed on seeds, taken on or near the ground. The *Passer* sparrows are rather noisy, using a variety of harsh, chirping notes; the others have more varied songs and rather harsh calls. **pp.212–215**

Wagtails and Pipits Subfamily Motacillinae

These are small, slender, terrestrial birds with long legs, relatively long toes and a thin, pointed bill. Most wagtails wag the tail up and down, and so do some pipits. Some wagtails exhibit wide geographical plumage variation. All walk with a deliberate gait and run rapidly. The flight is undulating and strong. They feed mainly by picking insects from the ground as they walk along, or by making short, rapid runs to capture insects they have flushed; they also catch prey in mid-air. Song flights are characteristic of many pipits. Both wagtails and pipits call in flight, and this is often a useful identification feature. They are usually found singly or in pairs in the breeding season, and in scattered flocks in autumn and winter. **pp.216–221**

Accentors Subfamily Prunellinae

Small, compact birds resembling *Passer* sparrows in appearance, but with a more slender and pointed bill. Accentors forage quietly and unobtrusively on the ground, moving by hopping or in a shuffling walk; some species also run. In summer, accentors are chiefly insectivorous, and in winter they feed mainly on seeds. Their flight is usually low over the ground and sustained over only short distances. **pp.222–223**

Weavers Subfamily *Ploceinae*

Small, rather plump, finch-like passerines with a large, conical bill. Adults feed chiefly on seeds and grain, supplemented by invertebrates; the young are often fed on invertebrates. Weavers inhabit grassland, marshes, cultivation and very open woodland. They are highly gregarious, roosting and nesting communally, and are noted for their elaborate roofed nests.

pp.224–225

Estrildine Finches Subfamily *Estrildinae*

Estreldine finches are small, slim passerines with a short, stout, conical beak. They feed chiefly on small seeds, which they pick up from the ground or gather by clinging to stems and pulling the seeds directly from seed-heads. Their gait is a hop or, occasionally, a walk. Outside the breeding season all species are gregarious. Their flight is fast and undulating.

pp.224–225

■ ■ Finches and Buntings Fringillidae ■ ■

Finches Subfamily *Fringillinae*

This subfamily contains small to medium-sized passerines that have a strong, conical bill used for eating seeds. They forage on the ground, though some species also feed on seed-heads of tall herbs, and blossoms or berries of bushes and trees. Finches are highly gregarious outside the breeding season, and their flight is fast and undulating.   **pp.224–231**

Buntings Subfamily *Emberizinae*

Buntings are small to medium-sized terrestrial passerines with a strong, conical bill designed for shelling seeds, usually of grasses; adults also eat insects in summer. They forage by hopping or creeping on the ground. Their flight is undulating. Buntings are usually gregarious outside the breeding season, feeding and roosting in flocks. They occur in a wide variety of open habitats.   **pp.230–233**

PARTRIDGES, QUAILS AND BUTTONQUAILS, PLATE 1

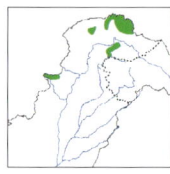

1 **SNOW PARTRIDGE** *Lerwa lerwa* 38 cm
ADULT Vermiculated dark brown and white upperparts, chestnut streaking on underparts, and red bill and legs. Often occurs in large parties, and can be very tame. Males advertise with low whistling calls, and noisy cackling when going to roost. High-altitude rocky and grassy slopes with scrub. Resident; 3300–5200 m. Rare in Hazara and uncommon in Gilgit and the Khunjerab.

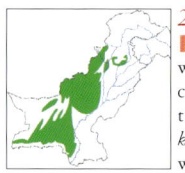

2 **SEE-SEE PARTRIDGE** *Ammoperdix griseogularis* 26 cm
a MALE and **b** FEMALE Rather uniform, sandy-coloured partridge. Male has white eye-stripe and chestnut and black flank stripes. Female has cream supercilium and throat, grey flecking on neck, and pinkish-buff and grey vermiculations on mantle and breast. Much smaller than Chukar. Males repeat inflected *khooit-khooit* call from prominent rock. When flushed, wings produce distinctive whirring. Dry rocky foothills, sand dunes and cultivation edges. Widely distributed west of the Indus, throughout Baluchistan from sea coast to northern foothills and across Punjab Salt Range.

3 **CHUKAR** *Alectoris chukar* 38 cm
ADULT Black gorget encircling throat, barring on flanks, and red bill and legs. Aggressive males call, a repeated *chuk-a-ka-chuk-a-ka* from a prominent rock. Easily trapped and popular cagebird. Open rocky or grassy hills; dry terraced cultivation. Resident. Widely distributed in mountainous tracts from central Baluchistan through NWFP, Gilgit, Hunza and Baltistan, up to 5000 m.

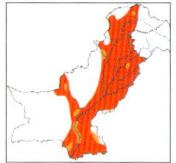

4 **COMMON QUAIL** *Coturnix coturnix* 20 cm
a MALE and **b** FEMALE Male has black 'anchor' mark on throat (which may be lacking), and buff or rufous breast with pale streaking. Female lacks 'anchor' mark and has blackish spotting on buffish breast. Call a liquid repeated *kwik-whi-kwik*. Often trapped for food. Crops and grassland. Passage migrant from April to May and from August to October, with some birds staying to breed in Chitral, Gilgit and Hunza.

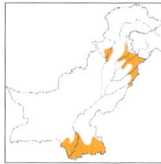

5 **RAIN QUAIL** *Coturnix coromandelica* 18 cm
a MALE and **b** FEMALE Male has strongly patterned head and neck, black on breast, and streaking on flanks. Female smaller than female Common, otherwise almost identical, but has unbarred primaries. Call a double whistle *which-which*, repeated 3–5 times. Crops, grassland, grass and scrub jungle. Localised monsoon visitor to southeast Sind and northeast Punjab, breeding after good rains.

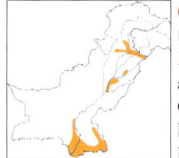

6 **SMALL BUTTONQUAIL** *Turnix sylvatica* 13 cm
MALE Very small size and pointed tail. Buff edges to scapulars form prominent lines, and rufous mantle and coverts are boldly fringed buff, creating scaly appearance. Underparts are similar to many Yellow-legged. Bill grey and legs pinkish. Only females call, as befits their polyandrous breeding: a surprisingly loud throbbing lasting 10 seconds, repeated at intervals. Tall grassland. Erratic summer breeding visitor to extreme southeastern Sind and northern parts of the Punjab Salt Range, westwards to adjacent foothills.

7 **YELLOW-LEGGED BUTTONQUAIL** *Turnix tanki* 15–16 cm
a MALE and **b** FEMALE Yellow legs and bill. Comparatively uniform upperparts (lacking scaly or striped appearance), and buff coverts with bold black spotting. Polyandrous breeding, with only the female calling; call is reminiscent of the distant chuffing of a two-stroke engine. Scrub and grassland, and crops. Uncommon visitor to lower Sind after good rains. More widespread and commoner in northern Punjab, westwards into Kohat.

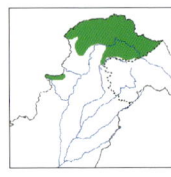

1 **HIMALAYAN SNOWCOCK** *Tetraogallus himalayensis* 72 cm
ADULT Very large, with chestnut neck stripes, whitish breast contrasting with dark grey underparts, and chestnut flank stripes. In flight, shows extensive white in primaries, little or no white in secondaries, and greyish rump. Call is a far-carrying *cour-lee-whi-whi*. High-altitude rocky slopes and alpine meadows. Resident, 3000–5000 m. Rare sedentary resident in Safed Koh, but widespread throughout northern areas from Chitral to Baltistan.

2 **GREY FRANCOLIN** *Francolinus pondicerianus* 33 cm
ADULT Buffish throat with fine dark necklace. Finely barred upperparts with shaft streaking, and finely barred underparts. Sexes similar, but female lacks spurs. Males call year-round, mostly at sunrise and sunset, a cheery repeated *kha-teeja-khateeja*. Common resident, adapted to semiarid foothills, irrigated forest plantations and uncultivated thorn scrub, up to 900 m.

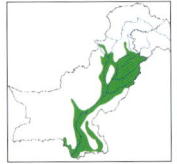

3 **BLACK FRANCOLIN** *Francolinus francolinus* 34 cm
a MALE and **b** FEMALE Male has black face with white ear-covert patch, rufous neck-band, and black underparts with white spotting. Female has rufous hind-neck, buffish supercilium and dark eye-stripe, streaked appearance to upperparts, and heavily barred or spotted underparts. Male usually calls from a vantage point: a staccato grating *chick-ghweek* followed by more drawn-out *gheek-ka-gheek*. Heavily persecuted both for food and as a popular cagebird. Resident, less common than Grey Francolin, confined to Indus basin and main tributaries up to Margalla Hills, preferring moister habitat.

4 **WESTERN TRAGOPAN**
Tragopan melanocephalus M 68–73 cm, F 60 cm
a MALE and **b** FEMALE Male has orange fore-neck, blackish underparts that are boldly spotted with white, red hind-neck, and red facial skin. Female is dull greyish brown in coloration, with white spotting on underparts. Call is a nasal, wailing *khuwaah*, repeated in bouts of 7–15 calls, likened to the wailing of a child or goat. Also a more abrupt *waa, waa, waa* when alarmed. Temperate and sub-alpine forest. Resident. Rare and with disjunct distribution in higher forested valleys of Indus Kohistan and Neelum valley. Summers up to 3350 m, descending to 2100 m in winter. Globally threatened (Vulnerable).

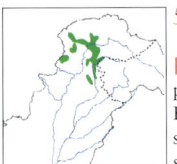

5 **KOKLASS PHEASANT**
Pucrasia macrolopha M 58–64 cm, F 52.5–56 cm
a MALE and **b** FEMALE Male has bottle-green head and ear-tufts, white neck-patch, chestnut on underparts, and streaked appearance to grey upperparts. Female has white throat, short buff ear-tufts, and heavily streaked body. Both sexes have wedge-shaped tail. Males start calling pre-dawn from roosting tree, a staccato grating *ka-kakaaah-kah*, repeated at intervals. Resident, 2100–2700 m. Relatively common and widespread wherever there are extensive patches of Himalayan temperate forest, from Murree Hills westwards to Chitral.

PHEASANTS, PLATE 3

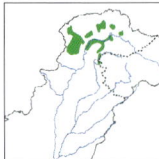

1 HIMALAYAN MONAL
Lophophorus impejanus M 70 cm, F 63.5 cm

a MALE and **b** FEMALE Male is iridescent green, copper and purple, with white patch on back and cinnamon-brown tail. Female has white throat, short crest, boldly streaked underparts, white crescent on uppertail-coverts, and narrow white tip to tail. When alarmed both sexes utter whistling *kleeh-wick-kleeh-wick* calls. In the breeding season males produce more melodious Eurasian Curlew-like whistles *kur-lieo-kleeh-kur-lieu-kleeh-kleeh-kurlieo-kurlieo*. Resident; recorded up to 4900 m. Roosts in Himalayan temperate forest, foraging by day, mostly on alpine grassy slopes. Thinly distributed from Chitral eastwards to the Kashmir 'cease-fire' line.

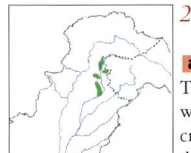

2 KALIJ PHEASANT
Lophura leucomelanos M 65–73 cm, F 50–60 cm

a MALE and **b** FEMALE Both sexes have red facial skin and down-curved tail. There is considerable racial variation in this species. The race occurring in the western Himalayas, *L. l. hamiltonii*, is distinctive: male has white or grey-brown crest, broad white barring on rump, and heavily scaled upperparts. Female is reddish brown, with greyish-buff fringes producing a scaly appearance. Males display by wing drumming, and produce rapid low-pitched clucks *terook-terook-terook*, ascending and accelerating into a squealing *ter-rr-eeoh-tuk-tuk tuk-treeh*. Females are also very vocal, with rapid clucks and squealing chirrups. Thinly distributed at lower elevations in predominantly deciduous subtropical Himalayan forest, 460–2100 m.

3 CHEER PHEASANT
Catreus wallichii M 90–118 cm, F 61–76 cm

a **c** MALE and **b** FEMALE Long, broadly barred tail, pronounced crest, and red facial skin. Male is more cleanly and strongly marked than female, with pronounced barring on mantle, unmarked neck, and broader barring across tail. Utters distinctive pre-dawn and dusk contact calls, including high-piercing whistles, *chewewoo*. Steep, craggy hillsides in mountainous regions with scrub and secondary growth. Current status uncertain; possibly survives in small numbers in scattered localities (e.g. in Neelum and Jhelum valleys) but may be extirpated. Globally threatened (Vulnerable).

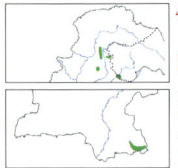

4 INDIAN PEAFOWL
Pavo cristatus M 180–230 cm, F 90–100 cm

a MALE and **b** FEMALE Male has blue neck and breast, and spectacular glossy green train of elongated uppertail-covert feathers with numerous ocelli. Female has whitish face and throat and white belly, and lacks elongated uppertail-coverts. Call is a trumpeting, far-carrying and mournful *kee-ow, kee-ow, kee-ow*. Dense riverine vegetation and open forest. A remnant population still survives in the border regions of Tharparkar, where the Hindu peasant population protects it. A small population struggles to survive in the foothill regions of Sialkot and westwards into Poonch. A well-known feral population survives at Kallar Kahar in the Punjab Salt Range.

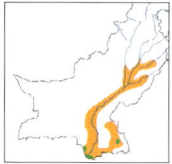

1 LESSER WHISTLING-DUCK *Dendrocygna javanica* 42 cm

a b ADULT Smaller than vagrant Fulvous Whistling-duck (see Appendix), and distinguished from that species by greyish-buff head and neck, dark brown crown, lack of well-defined dark line down hind-neck, bright chestnut patch on forewing, and chestnut uppertail-coverts. Both species of whistling-duck have a rather weak, deep-flapping flight, when they show dark upperwing and underwing. Mainly summer visitor, flocks arriving mid-May and showing a preference for reed-fringed pools and extensive rice-growing tracts. Small numbers remain year-round in lower Sind. Flocks utter rapid double-noted whistle, mainly in flight.

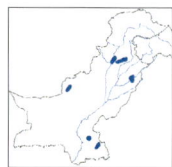

2 WHITE-HEADED DUCK *Oxyura leucocephala* 43–48 cm

a MALE and **b** FEMALE Swollen base to bill, and pointed tail, which is often held erect. Male has white head with black cap (bill becomes bright blue in breeding condition). Female has striped head. Winter visitor in small numbers to the Punjab Salt Range around the Khabbaki–Uchchali Lakes in the Soon Valley. Males occasionally give low grunting calls throughout the year and females respond with a goose-like *grek*. They undergo complete flightless moult while in Pakistan. Globally threatened (Endangered).

3 COMB DUCK *Sarkidiornis melanotos* 56–76 cm

a b MALE, **c** FEMALE and **d** JUVENILE Whitish head, speckled with black, and whitish underparts with incomplete narrow black breast-band. Upperwing and underwing blackish. Male has fleshy comb. Female lacks comb and is much smaller with duller upperparts. Pools in well-wooded country. Believed extirpated in Pakistan, with last reliable records from lower Sind in the early 1930s.

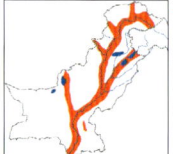

4 GREYLAG GOOSE *Anser anser* 75–90 cm

a b ADULT Large grey goose with pink bill and legs. Shows pale grey forewing in flight. Winter migrant entering Pakistan along the Indus and Chenab valleys. Most flocks continue southeastwards into India, with occasional flocks staying on the larger lakes or barrage headponds.

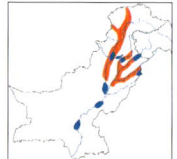

5 BAR-HEADED GOOSE *Anser indicus* 71–76 cm

a b ADULT and **c** JUVENILE Adult has white head with black banding, and white line down grey neck. Has black-tipped yellowish bill and yellowish legs. Juvenile has white face and dark grey crown and hind-neck. Plumage paler steel-grey, with paler grey forewing, compared with Greylag. In flight, their calls are deeper pitched than those of Greylags. Winter migrant, entering Pakistan along the Indus and its tributaries, and even less common than the Greylag. Main wintering population stays around Taunsa Barrage headpond and seepage zones, with few records from Sind lakes in recent years.

6 GREATER WHITE-FRONTED GOOSE *Anser albifrons* 66–86 cm

a b ADULT and **c** JUVENILE Adult has white band at front of head, black barring on belly, orange-pink bill, and orange legs and feet. Upperwing more uniform than in Greylag. Juvenile lacks white frontal band and belly barring. See Appendix for comparison with Lesser White-fronted Goose. Large rivers and lakes. Rare vagrant to Pakistan along the Indus river. No reliable records after 1968.

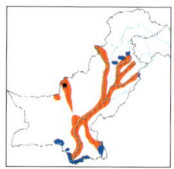

1 **COMMON SHELDUCK** *Tadorna tadorna* 58–67 cm
a b MALE, **c** FEMALE and **d e** JUVENILE Adult has greenish-black head and neck, and largely white body with chestnut breast-band and black scapular stripe. Female is very similar to male, but slightly duller and lacks knob on bill. Juvenile lacks breast-band and has sooty-brown upperparts. White upperwing- and underwing-coverts contrast with black remiges in flight in all plumages. Small numbers winter in remote brackish wetlands around the border of Badin District and along the sea coast near Karachi. Occasional small flocks occur on lakes in the Salt Range and Baluchistan.

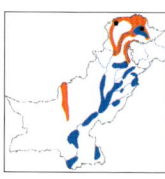

2 **RUDDY SHELDUCK** *Tadorna ferruginea* 61–67 cm
a b MALE and **c** FEMALE Rusty orange, with buffish head; white upperwing- and underwing-coverts contrast with black remiges in flight. Breeding male has black neck-band. Normally social, with males giving loud honking calls and females responding with deeper *aughaugh* calls. Frequent winter migrant to Indus plains, favouring brackish lakes and salt marshes, and always found in areas surrounded by open bare ground. Evidence of breeding around small lakes in Karumbar Valley and Chitral–Ishkuman boundary.

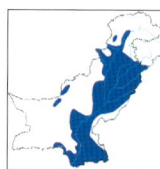

3 **GADWALL** *Anas strepera* 39–43 cm
a b MALE and **c d** FEMALE White patch on inner secondaries in all plumages. Male is mainly grey with white belly and dark patch at rear; bill is dark grey. Female similar to female Mallard, but has orange sides to dark bill and clear-cut white belly. Females often noisy with decrescendo quacking calls, while males respond with deeper croaking calls. Widespread winter visitor throughout Sind and Punjab wetlands, preferring less alkaline lakes with plenty of aquatic vegetation.

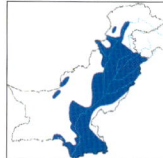

4 **EURASIAN WIGEON** *Anas penelope* 45–51 cm
a b MALE and **c d** FEMALE Male has yellow forehead and forecrown, chestnut head and pinkish breast; shows white forewing in flight. Female has rather uniform head, breast and flanks. (See Appendix for comparison with Falcated Duck.) In all plumages, shows white belly and rather pointed tail in flight. Very vocal, their characteristic *wheeoo* whistle frequently uttered. Abundant winter visitor, favouring larger, more open freshwater lakes, with smaller numbers wintering on salt marshes within the Indus delta.

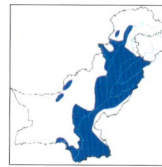

5 **MALLARD** *Anas platyrhynchos* 50–65 cm
a b MALE and **c d** FEMALE In all plumages, has white-bordered purplish speculum. Male has yellow bill, dark green head and purplish-chestnut breast. Female is pale brown and boldly patterned with dark brown; bill variable, patterned mainly in dull orange and dark brown. Reed-fringed lakes or seepage pools. Widespread winter visitor, although less common in Sind.

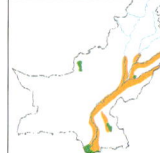

6 **SPOT-BILLED DUCK** *Anas poecilorhyncha* 58–63 cm
a b MALE and **c** FEMALE Yellow tip to bill, dark crown and eye-stripe, spotted breast and boldly scalloped flanks, and white tertials. Sexes similar, but male has red loral spot (brighter than illustrated) and is more strongly marked than female. Calls similar to those of Mallard, females uttering a rapid repeated *quack-quack* and males a more nasal wheezy *quack*. A summer monsoon visitor with a small resident population on lower Sind lakes. Breeding recorded in 1987 on Kushdil Khan Lake, Baluchistan.

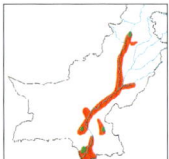

1 **COTTON PYGMY-GOOSE**
Nettapus coromandelianus 30–37 cm
a b MALE, **c** ECLIPSE MALE and **d e** FEMALE Small size, with greenish upperparts and whitish underparts. Male has broad white band across wing, and female has white trailing edge to wing. Male has white head and neck, black cap and black breast-band. Eclipse male and female have dark stripe through eye. Males utter rapid quacking *qua-qua ker-gah-qua-qua ger-gah* in display flights. Uncommon year-round resident, augmented by summer monsoon visitors, which have increased in numbers during the last 50 years. Confined mostly to lower Sind lakes, although it has spread up the Indus as far as the headpond of Jinnah Barrage at Mianwali.

2 **COMMON TEAL** *Anas crecca* 34–38 cm
a b MALE and **c d** FEMALE Male has chestnut head with green band behind eye, white stripe along scapulars, and yellowish patch on undertail-coverts. Female has rather uniform head, lacking pale loral spot of female Garganey. In flight, both sexes have broad white band along greater coverts, and green speculum with narrow white trailing edge; forewing of female is brown. Flocks of males call with inflected whistles *kerick-kerick*, females responding with low-pitched clucking. (See Appendix for comparison with Baikal Teal.) Very abundant winter visitor, preferring shallow tamarisk-dotted pools alongside larger water bodies.

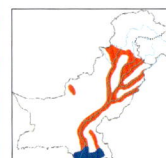

3 **GARGANEY** *Anas querquedula* 37–41 cm
a b MALE and **c d** FEMALE Male has white stripe behind eye, and brown breast contrasting with grey flanks; shows blue-grey forewing in flight. Female has more patterned head than female Common Teal, with more prominent supercilium, whitish loral spot, pale line below dark eye-stripe, and dark cheek-bar; shows pale grey forewing and broad white trailing edge to wing in flight. In spring, males are already displaying with head-jerking, wing-flapping and their peculiar clicking calls. Lakes, marshes and small ponds. Predominantly a spring and autumn passage migrant, when huge numbers pass through to and from the main wintering grounds in southern India and Sri Lanka. Small numbers are also present in winter.

4 **NORTHERN PINTAIL** *Anas acuta* 51–56 cm
a b MALE and **c d** FEMALE Long neck and pointed tail. Male has chocolate-brown head, with white stripe down sides of neck. Female has comparatively uniform buffish head and slender grey bill; in flight, shows combination of indistinct brownish speculum, prominent white trailing edge to secondaries, and greyish underwing. Common winter visitor, preferring large bodies of open water and often keeping to large same-sex flocks, which do not vocalise.

5 **NORTHERN SHOVELER** *Anas clypeata* 44–52 cm
a b MALE and **c d** FEMALE Long, spatulate bill. Male has dark green head, white breast, chestnut flanks and blue forewing. Female recalls female Mallard in plumage but has blue-grey forewing. Common winter visitor, being among the earliest and latest migrant ducks (late August to late May). They prefer muddy pools that are drying out and flooded grassland, but will spend the day on large lakes and rivers.

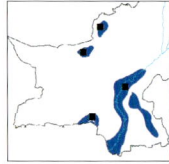

6 **MARBLED DUCK** *Marmaronetta angustirostris* 39–42 cm
a b ADULT A rather uform sandy-brown duck. Has shaggy hood, dark mask, and diffusely spotted body. Upperwing rather uniform and underwing very pale. Occurs in small numbers in winter, mostly on Sind lakes, with scattered breeding records from Zangi Nawar Lake, Baluchistan (up to 200 pairs depending on water levels), and small seepage lakes in Nawabshah and Sanghar districts of Sind. No recent wintering records from Punjab. Globally threatened (Vulnerable).

DIVING DUCKS, PLATE 7

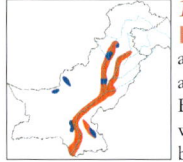

1 **RED-CRESTED POCHARD** *Netta rufina*   53–57 cm
a b MALE and **c d** FEMALE Large, with square-shaped head. Shape at rest and in flight more like dabbling duck. Male has rusty-orange head, black neck and breast, and white flanks. Female has pale cheeks contrasting with brown cap. Both sexes have largely white flight feathers on upperwing, and whitish underwing. Uncommon winter visitor from mid-October, preferring deeper water bodies, particularly irrigation barrage headponds and seepage lakes on Tharparkar border. Chashma Barrage on the Indus holds the largest concentration: up to 3000 some winters.

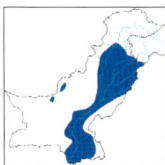

2 **COMMON POCHARD** *Aythya ferina*   42–49 cm
a b MALE, **c d** FEMALE and **e** IMMATURE MALE Large with domed head. Pale grey flight feathers and grey forewing result in different upperwing pattern from other *Aythya*. Male has chestnut head, black breast, and grey upperparts and flanks. Female has brownish head and breast contrasting with paler brownish-grey upperparts and flanks; lacks white undertail-coverts; eye is dark, and bill has grey central band. Common winter visitor, forming large flocks on open water bodies.

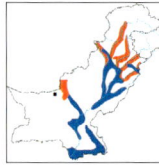

3 **FERRUGINOUS POCHARD** *Aythya nyroca*   38–42 cm
a b MALE and **c** FEMALE Smallest *Aythya* duck, with dome-shaped head. Chestnut head, breast and flanks and white undertail-coverts. Female is duller than male and has a dark iris. (See Appendix for comparison with Baer's Pochard.) In flight, shows extensive white wing-bar and white belly. Rare winter visitor, favouring secluded reed- and sedge-fringed pools. When water conditions are favourable, it breeds at Zangi Nawar Lake, Baluchistan (*c.*15 pairs).

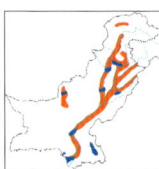

4 **TUFTED DUCK** *Aythya fuligula*   40–47 cm
a b MALE, **c** IMMATURE MALE, **d e** FEMALE and **f** FEMALE WITH SCAUP-LIKE HEAD Breeding male is glossy black, with prominent crest and white flanks. Eclipse/immature males are duller, with greyish flanks. Female is dusky brown, with paler flanks; some females may show white face-patch, recalling Greater Scaup (see Appendix), but they usually also show tufted nape and squarer head shape. Female has yellow iris. Common winter visitor, preferring large, open water bodies. Large concentrations at Khinjir Lake, Thatta District.

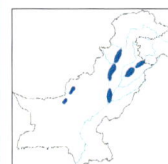

5 **COMMON GOLDENEYE** *Bucephala clangula*   42–50 cm
a b MALE and **c d** FEMALE Stocky, with bulbous head. Male has dark green head, with large white patch on lores. Female and immature male have brown head, indistinct whitish collar, and grey body, with white wing-patch usually visible at rest. Swims with body flattened, and partially spreads wings when diving. In flight, both sexes show distinctive white patterning on wing. Rare and occasional winter straggler, mainly on major rivers and Indus coastal creeks.

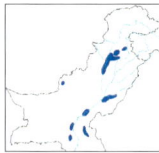

6 **SMEW** *Mergellus albellus*   38–44 cm
a b MALE and **c d** FEMALE Much smaller than the mergansers. Male is mainly white, with black markings. Immature male and female have chestnut cap, with white throat and lower ear-coverts. Occasional irruptions in severe winters, with flocks on Indus and Salt Range lakes, late January to early February.

7 **RED-BREASTED MERGANSER** *Mergus serrator*   52–58 cm
a b MALE and **c d** FEMALE Male has spiky crest, white collar, ginger breast and grey flanks. Female and immature male have chestnut head, which merges with grey of neck. Slimmer than Common Merganser, and with finer bill; chestnut of head and upper neck contrasts less with grey lower neck than in Common Merganser. In flight, white wing-patch is broken by black bar, unlike on Common. Rare, with small numbers wintering along the Arabian Sea coast of Pakistan.

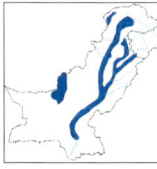

8 **COMMON MERGANSER** *Mergus merganser*   58–72 cm
a b MALE and **c d** FEMALE Male has dark green head and whitish breast and flanks (with variable pink wash). Female and immature male have chestnut head and greyish body. Rare winter visitor, mainly to Khushdil Khan Lake, Baluchistan, northern rivers of Punjab, and Rawal and Nammal lakes.

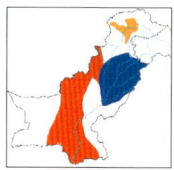

1 EURASIAN WRYNECK *Jynx torquilla*   16–17 cm
ADULT Cryptically patterned with grey, buff and brown. Has dark stripe down nape and mantle, and long, barred tail. In breeding season, males call from tree-tops, a rapidly repeated *quee-quee quee*. Thinly distributed winter visitor in the Punjab, and double passage migrant in lower Sind and central Baluchistan. Occasional breeding in northern areas.

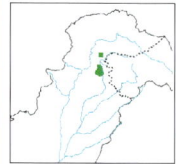

2 SPECKLED PICULET *Picumnus innominatus*   10 cm
a MALE and **b** FEMALE Tiny size. Greenish upperparts, whitish underparts with black spotting, black ear-covert patch and malar stripe, and white in black tail. Male has orange on forehead, which is lacking in female. Calls a rapidly repeated high-pitched *titi-ti*. Very rare disjunct population in temperate deciduous forest from Margalla Hills, westwards to Indus Kohistan.

3 GREY-CAPPED PYGMY WOODPECKER
Dendrocopos canicapillus   14 cm
a MALE and **b** FEMALE Very small. Has grey crown, blackish eye-stripe and indistinct blackish malar stripe. Upperparts blackish, barred with white. Underparts are dirty fulvous streaked with black, and lacks any rufous on vent. Calls a weak, rapid *tit-titerh-r-r-r-h*. Rare and irregular visitor to lower outer Himalayan foothills with predominantly deciduous trees.

4 BROWN-FRONTED WOODPECKER
Dendrocopos auriceps   19–20 cm
a MALE and **b** FEMALE Brownish forehead and forecrown, yellowish central crown, white barring to blackish upperparts, prominent black moustachial stripe, well-defined streaking on underparts, pink undertail-coverts, and unbarred central tail feathers. Display calls are a weak *chitter-chitter chiter-r-r-h*, and contact calls are a *chik-chik*. Small resident population largely confined to subtropical pine zone (*Pinus roxburghii*/*P. gerardiana*), from Waziristan across to outer Murree foothills.

5 FULVOUS-BREASTED WOODPECKER
Dendrocopos macei   18–19 cm
a MALE and **b** FEMALE White barring on black mantle and wing-coverts, and diffusely streaked buffish underparts. Male has red crown, which is black on female (compare with Brown-fronted). Weak drumming in breeding season along with *pik-pik* contact calls. Rare and erratically distributed at lower Himalayan elevations with dry deciduous scrub forest.

6 YELLOW-CROWNED WOODPECKER
Dendrocopos mahrattensis   17–18 cm
a MALE and **b** FEMALE Yellowish forehead and forecrown, white-spotted upperparts, poorly defined moustachial stripe, dirty underparts with heavy but diffuse streaking, red patch on lower belly, and bold white barring on central tail feathers. Relatively silent species. Contact calls are a *peek-peek* and occasional repeated *kik-kikkik-kir-r-r-h*. Uncommon and confined to lowland plains and semi-desert, wherever scattered trees occur, from Karachi up to Vale of Peshawar.

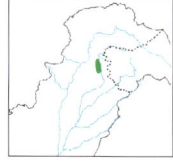

7 RUFOUS-BELLIED WOODPECKER
Dendrocopos hyperythrus   20 cm
a MALE and **b** FEMALE Whitish face and rufous underparts. Lacks white wing-patch. Male has red crown; female has black crown with white spots. Juvenile has barred underparts. Both sexes drum, and male's advertising calls are rapid and high-pitched. Scarce resident, confined to Murree Hills and Poonch, preferring lower Himalayan temperate forest with predominantly deciduous trees.

WOODPECKERS, PLATE 9

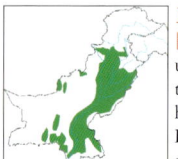

1 **SIND WOODPECKER** *Dendrocopos assimilis*   20–22 cm

a MALE and **b** FEMALE Black mantle, white wing-patch and unmarked underparts. Has black moustachial stripe joining hind-neck. Lacks black border to rear of ear-coverts of Himalayan; also has larger white wing-patch and forehead, whiter underparts, broader white barring on wings and paler pink vent. Both sexes drum and call *chirrir-rirh-rirrh*. Very widespread endemic species, adapted to dry tropical thorn scrub and wild olive/acacia scrub forest, from sea coast up to dry foothills in the north. Scattered populations through central Baluchistan.

2 **HIMALAYAN WOODPECKER**
Dendrocopos himalayensis   23–25 cm

a MALE and **b** FEMALE Black mantle, white wing-patch and unmarked underparts. Has black rear border to ear-coverts. Both sexes drum and males give rapid chattering calls. Common in northern Himalayan forest, 2000–3200 m, with little altitudinal migration even in the hardest winters.

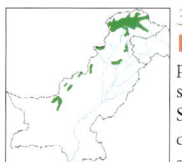

3 **SCALY-BELLIED WOODPECKER** *Picus squamatus*   35 cm

a MALE and **b** FEMALE Large greenish woodpecker, with scaling on underparts. Male has red crown. Female has a blackish crown marked with white. Both sexes have white supercilium and blackish moustachial stripe. Larger than similar Streak-throated Woodpecker (N.B. this species has not been recorded in Pakistan); differences from that species are pale bill and reddish eye, prominent moustachial stripe, unstreaked throat and upper breast (although these parts are streaked in juvenile), and barred tail. Advertising calls are a two-noted *klee-gah-kleee-gah* and contact calls are a *kuik-kuik*. Widespread resident, eclectic in its choice of habitat, from dry bush-studded stream beds to juniper forest, Himalayan temperate forest and groves of willow and poplar in far northern regions, 600–3000 m.

4 **GREY-HEADED WOODPECKER** *Picus canus*   32 cm

a MALE and **b** FEMALE Greenish-coloured woodpecker. Distinguished from Scaly-bellied by plain grey face and uniform greyish-green underparts. Male has scarlet forecrown contrasting with black nape. In female, crown and nape are blackish, marked with white. Calls are similar to those of Scaly-bellied Woodpecker. Very restricted in distribution, and confined to southern part of the Murree Hill range and probably declining due to deforestation on these lower slopes.

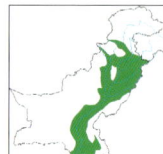

5 **BLACK-RUMPED FLAMEBACK**
Dinopium benghalense   26–29 cm

a MALE and **b** FEMALE Has brilliant golden-coloured upperparts, and is the only flameback species recorded in Pakistan. Distinguishing features from other flamebacks include smaller size, different head pattern (combination of black eye-stripe and throat, but lacking dark moustachial stripe), spotting on wing-coverts, and black rump. Calls are a whinnying, rapid *kyi-kyi-kyi*, not unlike those of White-throated Kingfisher. Common resident throughout Indus plains, especially favouring canal and roadside tree plantations, and older gardens. Less adapted to dry thorn scrub than Sind Woodpecker.

BARBETS, GREY HORNBILL, HOOPOE AND ROLLERS, PLATE 10

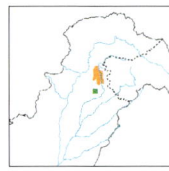

1 **GREAT BARBET** *Megalaima virens*   33 cm
ADULT Largest barbet, with large yellow bill, bluish head, brown breast and mantle, olive-streaked yellowish underparts, and red undertail-coverts. Has a monotonous, incessant and far-reaching *piho, piho*, uttered throughout the day; a repetitive *tuk tuk tuk* is often uttered in a duet, presumably by female. Confined to the Murree Hills in temperate mixed Himalayan forest. Some altitudinal migration during winter snows. Absent from Hazara District.

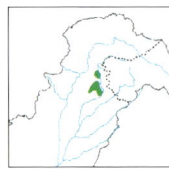

2 **BLUE-THROATED BARBET** *Megalaima asiatica*   23 cm
ADULT Medium-sized barbet with blue 'face' and throat, red forehead and hind-crown, and black band across crown. Juvenile has duller head pattern. Incessant calling throughout day, a rapid *chukaroo-chukaroo*. Resident in outer Murree foothills and tropical pine zone.

3 **COPPERSMITH BARBET** *Megalaima haemacephala*   17 cm
a ADULT and **b** JUVENILE Small barbet with crimson forehead and breast-patch, yellow patches above and below eye, yellow throat, and streaked underparts. Juvenile lacks red on head and breast. Largely silent in winter, calling incessantly from early March. Voice is a loud, metallic, repetitive and monotonous *tuk, tuk, tuk*. Open wooded country, groves and trees in cultivation and gardens. Resident throughout Indus plains from northern Sind up to Margalla Hills.

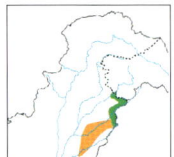

4 **INDIAN GREY HORNBILL** *Ocyceros birostris*   50 cm
a b MALE and **c** FEMALE Small hornbill with sandy brownish-grey upperparts, long tail that has a dark subterminal band and elongated central feathers, and white trailing edge to wing. Prominent black casque and extensive black at base of bill. Female similar to male, but with smaller bill and casque. Territorial call is a loud cackling and squealing *k-k-k-ka-e* or rapid piping *pi-pi-pi-pi-pipipieu* etc.; normal contact call is a kite-like *chee-ooww*. Open broadleaved forest, groves and gardens with fruiting trees. Scarce summer breeding visitor to northeast Punjab, with some pairs resident around Lahore and Kasur.

5 **COMMON HOOPOE** *Upupa epops*   31 cm
a b ADULT Orange-buff, with broad black-and-white wings, white-banded black tail and black-tipped fan-like crest. Frequently raises crest, especially when alarmed. Wings are broad and rounded, making it rather butterfly-like in flight. Voice is a repetitive and mellow *poop, poop, poop*, similar to that of Oriental Cuckoo. Open country, cultivation and villages. Common resident in northern Punjab, wintering in Sind and breeding in summer in northern areas, with population augmented by spring migrants from East Africa.

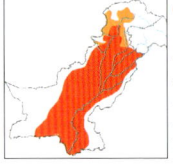

6 **EUROPEAN ROLLER** *Coracias garrulus*   31 cm
a b ADULT and **c** JUVENILE Turquoise head and underparts, and rufous-cinnamon mantle. Has black flight feathers and tail corners. Juvenile is much duller, and has whitish streaking on throat and breast; patterning of wings and tail helps separate it from Indian Roller. Open woodland and cultivation. Double passage migrant through Baluchistan northwards in spring and throughout Indus plains in autumn. Summer breeder in broader valleys of northern areas.

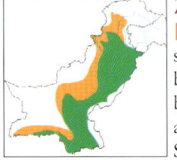

7 **INDIAN ROLLER** *Coracias benghalensis*   33 cm
a b ADULT and **c** JUVENILE Has rufous-brown on nape and underparts, white streaking on ear-coverts and throat, and greenish mantle. In flight, shows turquoise band across primaries and dark blue terminal band to tail. Juvenile is similar to adult but duller, with more heavily streaked throat and breast. Harsh grating calls rise to a crescendo. Cultivation, open woodland, groves and gardens. Common throughout Sind, Punjab and NWFP, avoiding extensive desert and higher hill tracts.

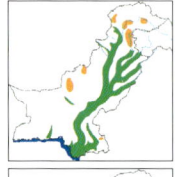

1 COMMON KINGFISHER *Alcedo atthis*   16 cm
ADULT Small kingfisher with turquoise upperparts and orange underparts. Flashes brilliant turquoise line down back in flight. High-pitched *chee* given in flight. Uncommon resident throughout Pakistan, preferring rivers and major wetlands, extending across far northern areas as summer breeder. Some local migration to sea coast in winter.

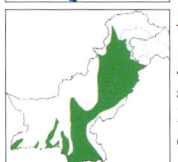

2 WHITE-THROATED KINGFISHER *Halcyon smyrnensis*   28 cm
ADULT White throat and centre of breast, brown head and most of underparts, and turquoise upperparts. Shows prominent white wing-patch in flight. Call is a loud rattling laugh; song is drawn-out musical whistle, *kililili*. Cultivation, forest edges, gardens and wetlands. Common, widespread and sedentary throughout Sind, Punjab and NWFP, extending into broader northern mountain valleys.

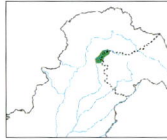

3 CRESTED KINGFISHER *Megaceryle lugubris*   41 cm
ADULT Much larger than Pied, with evenly barred wings and tail. Lacks supercilium, and has spotted breast, which is sometimes mixed with rufous. Striking crest is often held erect when alarmed. Rare resident and confined to the Neelum river in Azad Kashmir, with occasional pairs descending to smaller streams around Islamabad.

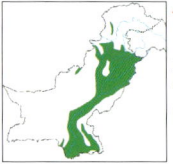

4 PIED KINGFISHER *Ceryle rudis*   31 cm
a MALE and **b** FEMALE Smaller than Crested, with white supercilium, white patches on wings, and black band(s) across breast. Female has single breast-band (double in male). Abundant sedentary species throughout wetlands of Punjab, and especially lower Sind. Penetrating broader valleys into Swat and Kohat.

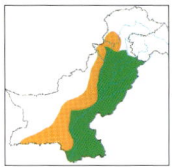

5 GREEN BEE-EATER *Merops orientalis*   16–18 cm
a b ADULT and **c** JUVENILE Small bee-eater, with blue cheeks, black gorget, and golden coloration to crown. Green tail with elongated central feathers. Juvenile has green crown and nape, yellowish throat; it lacks black gorget and has square-ended tail. Open country with scattered trees. Abundant throughout plains of Sind, Punjab and NWFP, avoiding mountainous areas, and largely absent from Baluchistan. Locally migratory to warmer areas in winter.

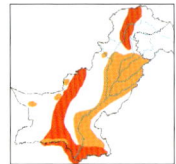

6 BLUE-CHEEKED BEE-EATER *Merops persicus*   24–26 cm
a b ADULT Larger than Green Bee-eater, with chestnut lower throat, whitish forehead and turquoise-and-white supercilium. Green upperparts, underparts and tail, although may show touch of turquoise on rump, belly and tail-coverts. See below for differences from Blue-tailed. Summer breeding migrant from East Africa, preferring swampy areas. Autumn passage of birds breeding in central Asia.

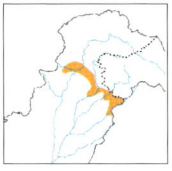

7 BLUE-TAILED BEE-EATER *Merops philippinus*   23–26 cm
a b ADULT and **c** JUVENILE Larger than Green Bee-eater, with chestnut throat, green crown and nape, and blue tail. Compared with Blue-cheeked, forehead and supercilium are mainly green, sometimes with a touch of blue on supercilium; chestnut of throat is more extensive and extends onto ear-coverts, and blue streak below mask is less extensive. Upperparts and underparts are washed with brown and blue (greener in Blue-cheeked). Juvenile is like washed-out version of adult; lacks elongated central tail feathers. Uncommon summer migrant, breeding along the foothill zone of northern Punjab.

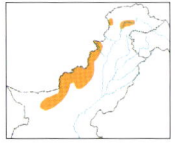

8 EUROPEAN BEE-EATER *Merops apiaster*   23–25 cm
a b c ADULT and **d** JUVENILE Yellow throat, black gorget, blue underparts, chestnut crown and mantle, and golden-yellow scapulars. Juvenile is duller than adult, but still shows chestnut on crown and well-defined yellowish throat. Widespread summer breeding visitor in open country, confined to western fringes of Baluchistan, NWFP and into broader valleys of northern areas.

CUCKOOS, PLATE 12

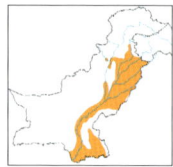

1 PIED CUCKOO *Clamator jacobinus*   33 cm
ADULT Adult is black and white with prominent crest. Has white patch at base of primaries. Crest smaller, upperparts browner and underparts more buffish on juvenile. Call is metallic, pleasant *piu…piu…pee-pee piu, pee-pee piu*. Broadleaved forest and well-wooded areas. A monsoon-season breeding visitor, fostering on Common and Jungle babblers, and confined to the Indus plains and Punjab Salt Range; does not penetrate the mountains.

2 COMMON HAWK CUCKOO *Hierococcyx varius*   34 cm
a ADULT and **b** JUVENILE Superficially resembles a *Cuculus* cuckoo but with broad banding on tail, rufous coloration to breast and whitish throat. Smaller than Large Hawk Cuckoo (see Appendix), with grey upperparts, more rufous on underparts, indistinct barring on belly and flanks, and narrow tail-banding. Juvenile has spotted flanks and narrow tail-banding. Calls incessantly, often at night, 3- or 4-noted *whee-pi-whit* or *whee-whee h'yar-ho*. Summer breeding migrant from March to late October. Confined to well-wooded country in northeastern plains and foothills; absent from Sind and Baluchistan.

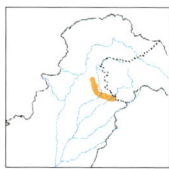

3 INDIAN CUCKOO *Cuculus micropterus*   33 cm
a MALE and **b** JUVENILE Brown coloration to grey upperparts and tail, broad barring on underparts, and pronounced white barring and spotting on tail. Juvenile has broad and irregular white tips to feathers of crown and nape, and white tips to scapulars and wing-coverts. Call is a descending, 4-noted whistle, *kwer-kwah…kwah-kurh*. Rare visitor to better wooded tracts of extreme northwest Punjab during monsoon.

4 EURASIAN CUCKOO *Cuculus canorus*   32–34 cm
a MALE and **b** HEPATIC FEMALE Finer barring on whiter underparts than Oriental. Male's call is a loud repetitive *cuck-oo…cuck-oo*; female has bubbling call. Widespread breeding visitor to higher mountainous tracts in Baluchistan and northern areas up to the alpine slopes of Pakistan's northern borders. Scattered records of arrivals in April in the plains, where they don't linger.

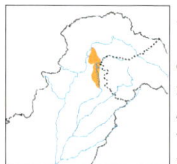

5 ORIENTAL CUCKOO *Cuculus saturatus*   30–32 cm
a MALE and **b** HEPATIC FEMALE Broader barring on buffish-white underparts compared with Eurasian; upperparts are a shade darker; head is paler. Hepatic female is slightly more heavily barred than Eurasian. Call is a resonant *ho…ho…ho…ho*, easily confused with the call of Common Hoopoe. Summer breeding visitor to forested areas from Murree Hills to western Hazara District. Often sympatric with Eurasian Cuckoo. Fosters its comparatively small eggs on *Phylloscopus* and *Seicercus* spp. in Pakistan.

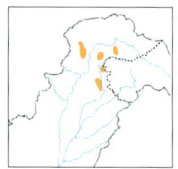

6 LESSER CUCKOO *Cuculus poliocephalus*   25 cm
a MALE and **b** HEPATIC FEMALE Smaller than Oriental; hepatic female can be bright rufous and indistinctly barred on crown and nape. Call is a strong, cheerful *pretty-peel-lay-ka-beet*. Uncommon summer breeding visitor to forest and well-wooded areas of northern Himalayas. Does not penetrate above tree-line at 3000 m. Believed to foster its eggs on *Cettia* warblers in Pakistan.

CUCKOOS, COUCALS AND PARAKEETS, PLATE 13

1 GREY-BELLIED CUCKOO *Cacomantis passerinus* 23 cm
a ADULT, **b** HEPATIC FEMALE and **c** JUVENILE A small cuckoo. Adult is grey with white vent and undertail-coverts. On hepatic female, underparts are whitish with dark barring, upperparts are bright rufous, with crown and nape only sparsely barred, and rufous tail is unbarred. Juvenile is either grey, with pale barring on underparts, or similar to hepatic female, or intermediate. Song is a clear, interrogative *pee-pipee-pee…pipee-pee*, ascending in scale and pitch; also a single plaintive repeated whistle. Late summer breeding migrant, confined to lower scrub-covered hills between 1500 m and 1800 m in Punjab and western Hazara District.

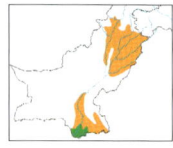

2 ASIAN KOEL *Eudynamys scolopacea* 43 cm
a MALE and **b** FEMALE Male is greenish black, with green bill. Female is spotted and barred with white. Song is a loud, rising, repeated *ko-el…ko-el…ko-el…ko-el*. Open woodland, gardens and cultivation. Resident in lower Sind, and summer migrant to Indus plains further north and westwards into better wooded parts of Kohat and Bannu. Fosters its eggs on House Crow, frequently tolerating its siblings.

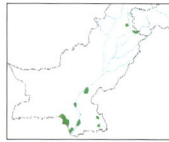

3 SIRKEER MALKOHA *Phaenicophaeus leschenaultii* 42 cm
ADULT Sandy coloration, yellow-tipped red bill, dark mask with white border, and bold white tips to tail. Rare sedentary resident, confined to northeast Punjab adjacent to foothills and, very rarely, in Sind, favouring extensive tropical thorn scrub.

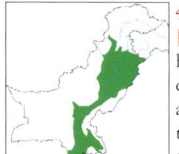

4 GREATER COUCAL *Centropus sinensis* 48 cm
a ADULT and **b** JUVENILE Large and black with chestnut wings. Juvenile is heavily barred. Call is a deep primate-like *hoop-hoop-hoop-hoop-hoop-hoop-hoop*, descending and then rising towards the end of the series. Favours tall grassland and thickets near cultivation, in particular areas of dense cover near major irrigation works. Common sedentary resident in Sind and Punjab, and in irrigated cultivation in NWFP. Absent from Baluchistan.

5 ALEXANDRINE PARAKEET *Psittacula eupatria* 53 cm
a MALE and **b** FEMALE Very large, with maroon shoulder-patch. Male has black chin-stripe and pink collar. Call is a loud, guttural *keeak* or *kee-ah*, much deeper than that of Rose-ringed. Locally common sedentary resident, favouring old gardens and irrigation colonies, into tropical pine zone. Absent in Baluchistan, extensive desert regions and Punjab Salt Range.

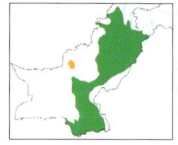

6 ROSE-RINGED PARAKEET *Psittacula krameri* 42 cm
a MALE and **b** FEMALE Green head and blue-green tip to tail. Male has black chin-stripe and pink collar. Call is a loud, shrill *kee-ah*. A serious pest of fruit crops. Abundant resident throughout Indus basin, favouring human settlements, even Quetta in sub-zero winters.

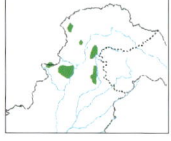

7 SLATY-HEADED PARAKEET *Psittacula himalayana* 41 cm
a MALE, **b** FEMALE and **c** IMMATURE Dark grey head, dark green upperparts, red-and-yellow bill and yellow-tipped tail. Female lacks maroon shoulder-patch of male. Immature lacks slaty head. Calls a strident *tree-tree*. Locally common in Himalayan mixed deciduous/coniferous forest. Breeds at 2100–2450 m and is an altitudinal migrant to foothills in winter.

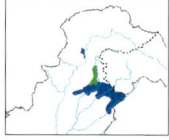

8 PLUM-HEADED PARAKEET *Psittacula cyanocephala* 36 cm
a MALE, **b** FEMALE and **c** IMMATURE Head is plum-red on male, pale grey on female. Yellow upper mandible, and white-tipped blue-green tail. Head of female is paler grey than Slaty-headed, and lacks black chin-stripe and half-collar of that species; shows yellow collar and upper breast. Call is higher pitched than Slaty-headed's, a distinctive *toowinck-toowinck*. Broadleaved forest and well-wooded areas. Occurs in Murree Hills and Poonch in the restricted zone east of Murree town, wintering in adjacent plains.

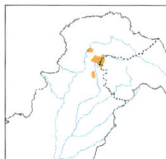

1 WHITE-THROATED NEEDLETAIL
Hirundapus caudacutus   20 cm

a b ADULT Large swift, with very fast and powerful flight. Has striking white throat and white horseshoe crescent on underparts. Also shows prominent pale 'saddle' on upperparts, and white patch on tertials. Flight calls are a rapid chitter. Uncommon summer visitor, favouring mountain summit ridges and peaks. Indus Kohistan and Hazara District east to Dunga Gali. No evidence of successful breeding in the subcontinent.

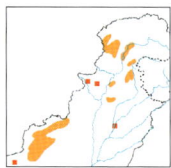

2 ALPINE SWIFT *Tachymarptis melba*   22 cm

a b ADULT Large, powerful swift with white throat, brown breast-band and white patch on belly. Calls are short noisy bursts of chittering. Colonially breeding migrant in suitable cliff faces, in main Himalayan valleys and northern Baluchistan. Favours lower altitudes than Common Swift. Birds on passage observed throughout Indus plains and Arabian Sea coast, and believed to winter in Africa.

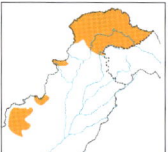

3 COMMON SWIFT *Apus apus*   17 cm

a b ADULT Uniform dark brown swift, with white throat; lacks white rump. Has prominently forked tail. Wheeling flocks give high-pitched screams. Widespread, common summer breeding visitor, extending throughout drier mountain regions of the north up to the Wakhan and east to Hunza. Less common in north-western Baluchistan. Absent from Murree Hills, but common further west in Kunhar Valley.

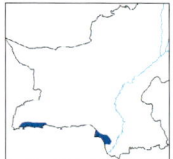

4 PALLID SWIFT *Apus pallidus*   17 cm

a b ADULT Superficially similar to Common Swift but with subtle differences, although good light conditions are needed for confident identification. Has pale grey-brown upperparts and underparts with darker eye-patch, and more extensive pale throat than Common. Shows contrast between dark outer primaries and inner wing-coverts and paler rest of wing. Underparts more distinctly scaled than in Common, and has dark-saddled, pale-headed appearance. Flight calls lower pitched than those of Common Swift. Rare and locally occurring winter visitor in suitable sea-cliff areas along the Baluchistan sea coast.

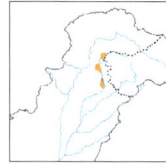

5 FORK-TAILED SWIFT *Apus pacificus*   15–18 cm

a b ADULT Blackish swift with white rump and indistinct white scaling on underparts. Slimmer bodied and longer winged than House Swift, with deeply forked tail. Flight calls are a loud, rapid *chik-chikchik*. A rare summer breeding visitor, mainly to northern Neelum Valley; favours forested slopes. Extirpated in the Murree Hills, where breeding was recorded at beginning of 20th century.

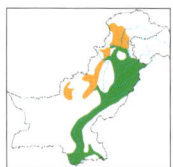

6 HOUSE SWIFT *Apus affinis*   15 cm

a b ADULT Small blackish swift with broad white rump-band. Compared with Fork-tailed, it is rather stocky, with comparatively short wings, and the tail appears square-ended or has only a slight fork. Flight calls are a high-pitched chittering. Conspicuous colonial resident, breeding throughout the Indus plains, with summer migrant breeders from northern Baluchistan up to Chitral. Favours crevices under old buildings in towns, and natural rock fissures, even under culverts in northern areas, for nesting.

SMALL OWLS, PLATE 15

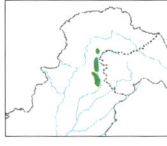

1 **MOUNTAIN SCOPS OWL** *Otus spilocephalus* 20 cm
ADULT Very small ear-tufts. Upperparts mottled with buff and brown; underparts indistinctly spotted with buff and barred with brown. Occurs as grey or fulvous-brown morphs. Call is a double whistle, *toot-too*. Common resident in restricted range of Himalayan mixed temperate forest from Murree Hills to eastern Hazara District, between 1800 m and 3000 m.

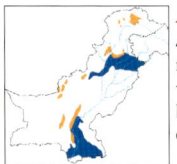

2 **PALLID SCOPS OWL** *Otus brucei* 22 cm
ADULT Compared with grey morph Oriental Scops, is paler, greyer and more uniform. Has less distinct scapular spots, and narrower dark streaking on underparts, which lack pale horizontal panels. Call is a resonant *whoop-whoop-whoop....* Scarce but widespread summer breeding visitor to drier mountainous regions, from Gilgit south to Sind Kohistan. Often sympatric with Eurasian and Oriental scops.

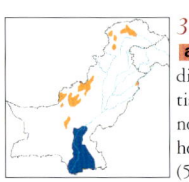

3 **EURASIAN SCOPS OWL** *Otus scops* 19 cm
a ADULT BROWN MORPH and **b** ADULT GREY MORPH From Oriental Scops, by different call and longer primary projection (6 or 7 primaries extending beyond tertials, 4–5 in Oriental). Browner or darker grey than Pallid Scops, with more pronounced darkening around eyes, more prominent white spots on scapulars, pale horizontal bars across streaked underparts, and more numerous pale barring on tail (5–7 bars compared to 2–4 in Pallid). Call a repeated single *took*. Common summer breeding visitor to higher mountain ranges in central Baluchistan, northern Chitral, Swat and Gilgit.

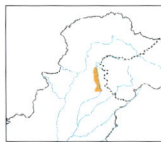

4 **ORIENTAL SCOPS OWL** *Otus sunia* 19 cm
a ADULT RUFOUS MORPH and **b** ADULT BROWN MORPH Prominent white scapular spots, streaked underparts and upperparts; lacks prominent nuchal collar. Iris yellow. Call is a frog-like *wut-chu-chraaii*. Very restricted uncommon summer visitor, confined to southern part of Murree Hill range, down to 600 m. Often sympatric with Mountain Scops.

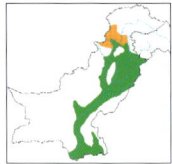

5 **COLLARED SCOPS OWL** *Otus bakkamoena* 23–25 cm
a ADULT GREY MORPH and **b** ADULT RUFOUS-BROWN MORPH Larger than other scops owls, with buff nuchal collar, finely streaked underparts, and indistinct buffish scapular spots. Iris dark orange or brown. Call is a subdued, frog-like *whuk*, repeated at irregular intervals. Common resident in the plains, preferring old groves of trees and forest plantations. Uncommon summer visitor to Murree Hills.

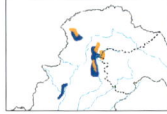

6 **COLLARED OWLET** *Glaucidium brodiei* 17 cm
ADULT Very small and heavily barred. Spotted crown, streaking on flanks, and owl-face pattern on upper mantle. Often hunts diurnally. Call is a 4-noted *poop-pu-pu-poop*, monotonously repeated. Common resident between 1800 m and 2750 m in Himalayan mixed coniferous/deciduous forest. Some altitudinal migration in winter.

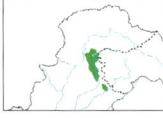

7 **ASIAN BARRED OWLET** *Glaucidium cuculoides* 23 cm
ADULT Heavily barred upperparts and finely barred wings and tail. Much larger than Collared Owlet. Mainly diurnal, and often perches prominently during the day. Call is a rapid ululating warble. Resident in lower Himalayan foothills at 460–1800 m from eastern Hazara District south to Mangla Dam.

8 **LITTLE OWL** *Athene noctua* 23 cm
ADULT Sandy brown, with streaked breast and flanks, and streaked crown. Crepuscular and partly diurnal; often perches prominently in daylight. Confined to western Baluchistan, favouring earth cliffs and rocky canyons, but confusingly sympatric with Spotted Owlet.

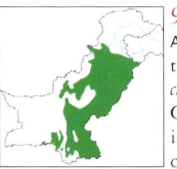

9 **SPOTTED OWLET** *Athene brama* 21 cm
ADULT White spotting on upperparts, including crown, and diffuse brown spotting on underparts. Mainly crepuscular and nocturnal. Call is a harsh, screechy *chirurr-chirurr-chirurr...*, followed by/alternated with *cheevak, cheevak, cheevak*. Common and widespread sedentary resident throughout the Indus basin, extending into extensive desert and dry foothill tracts in the north. Particularly favours canal and roadside tree plantations.

LARGE AND MEDIUM-SIZED OWLS, PLATE 16

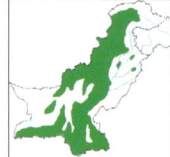

1 EURASIAN EAGLE OWL *Bubo bubo*   56–66 cm
a ADULT *B. b. hemachalana* and **b** ADULT *B. b. bengalensis* Very large owl. Upperparts mottled dark brown and tawny buff; underparts heavily streaked. Himalayan *hemachalana* is larger and paler than *bengalensis* of the lowlands and hills. Ear-tufts prominent and held erect when alert. Eyes are orange. Mainly nocturnal, but usually perches before sunset, and after sunrise, in a prominent position on cliff or rock. Call is a resonant *tu-whooh*. Resident, thinly distributed throughout drier mountain regions, avoiding coniferous forest and irrigated cultivation. Spreading eastwards into Thar Desert.

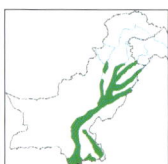

2 DUSKY EAGLE OWL *Bubo coromandus*   58 cm
ADULT Large grey owl. Upperparts greyish brown with fine whitish vermiculations and diffuse darker brown streaking. Underparts greyish white with brown streaking. Ear-tufts are prominent and held erect when alert. Eyes are golden yellow. Pairs duet noisily in breeding season, *whruck-whruk-wruk-uk-uk-yk-k-k-k*, accelerando. Uncommon resident, confined to the vicinity of the Indus and its main tributaries. Prefers well-wooded areas.

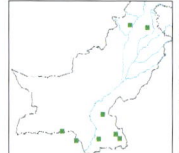

3 BROWN FISH OWL *Ketupa zeylonensis*   56 cm
ADULT Upperparts brown with dark streaking; underparts dull buff with clear, fine, dark streaking. Lacks dark border to facial discs as in Eurasian and Dusky eagle owls. Ear-tufts are prominent and held horizontally. Eyes are yellow. The legs and feet are bare, lacking feathering as in Eurasian and Dusky eagle owls. Partly diurnal, emerging from roost long before sunset; sometimes hunts during the day. Calls include a soft, deep *hup-hup-hu* and a wild *hu-hu-hu-hu…hu ha*. Extremely rare resident, confined to water with a good supply of fish, and adjacent tree cover. Has been observed hunting crabs along the sea coast near the Hab river.

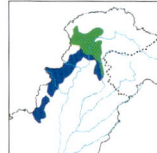

4 TAWNY OWL *Strix aluco*   45–47 cm
a ADULT *S. a. nivicola* and **b** *S. a. biddulphi* Rufous to dark brown, with heavily streaked underparts, white markings on scapulars, dark centre to crown, and pale forecrown stripes. The larger and greyer form *biddulphi* occurs in the northwest of the region. Nocturnal. Call is a *too-tu-whoo*, sometimes shortened to a 2-noted *turr-whooh*; also a low-pitched *ku-wack-ku-wack*. Resident in forested areas from western NWFP, Chitral, Swat and south through Hazara and the Murree Hills. Some altitudinal migration in winter.

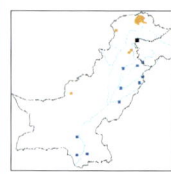

5 LONG-EARED OWL *Asio otus*   35–37 cm
a b ADULT Long ear-tufts, orange-brown facial discs and orange eyes. Additional differences from Short-eared are more heavily streaked belly and flanks, orange-buff base coloration to primaries and tail feathers, and lack of white trailing edge to wing. Winter visitor to plains, favouring uncultivated tracts, often roosting in plantations. Mainly migrant from central Asia, with occasional breeding in Himalayan coniferous forest. Colonial roosting recorded in Lal Sohanran. On passage and in winter it frequents tamarisk scrub and poplar plantations.

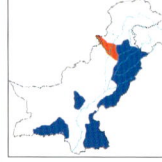

6 SHORT-EARED OWL *Asio flammeus*   37–39 cm
a b ADULT Streaked underparts and short ear-tufts. Buffish facial discs and yellow eyes. In flight, rather long and narrow wings show buffish patch at base of primaries and dark carpal patches. Often seen hunting during the day, quartering low over the ground in a leisurely manner, when it often hovers or glides with wings in a V. Uncommon, sparsely distributed winter visitor, favouring open treeless tracts, especially grassland on margins of larger wetlands.

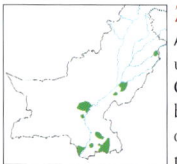

7 BARN OWL *Tyto alba*   36 cm
ADULT Unmarked white face, whitish underparts, and golden-buff and grey upperparts. Eyes are dark. Utters a variety of eerie screeching and hissing noises. Generally nocturnal. Usually hunts in flight, quartering the ground and often banking or hovering to locate prey. Uncommon resident, favouring vicinity of older towns with commensal rodents. Absent from mountainous areas and more plentiful around poultry-farming complexes in lower Sind.

NIGHTJARS, PLATE 17

1 GREY NIGHTJAR *Caprimulgus indicus*   27–32 cm

a b MALE and **c** FEMALE Greyer in overall coloration compared with Large-tailed, and less strikingly patterned (absent or less pronounced pale edges to scapulars, and less prominent pale tips to coverts). (See also Table 1 on p.241.) Song is a loud, resonant *chunk-chunk-chunk-chunk*.... Calls similar but quicker paced than those of Large-tailed. Summer breeding visitor, confined to Murree Hills and Agrore Valley in eastern Hazara District. Arrivals in foothills start calling when sympatric with Large-tailed and Savanna, but breeding occurs at 1800–2400 m in Himalayan mixed temperate forest.

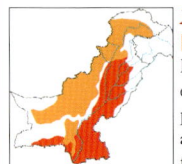

2 EURASIAN NIGHTJAR *Caprimulgus europaeus*   25 cm

a b MALE and **c** FEMALE Medium-sized grey nightjar with regular, bold lanceolate streaking on crown, nape and scapulars (latter with buffish outer edges). Similar in coloration to Grey Nightjar but more cleanly streaked and with pronounced pale edges to scapulars, and more prominent pale tips to coverts. (See also Table 1 on p.241.) Song is a continuous churring, plus a soft *quoit quoit* in flight. Summer breeding visitor to low rocky valleys and hills west of the Indus, from Karachi environs north to Chitral and eastwards to Deosai Plateau. Population winters mainly in East Africa, but there are occasional wintering records throughout the Indus plains.

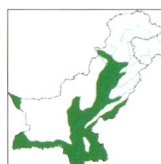

3 SYKES'S NIGHTJAR *Caprimulgus mahrattensis*   23 cm

a b MALE and **c** FEMALE Small grey or sandy-coloured nightjar. Rather uniform in appearance, with finely streaked crown, irregular black marks on scapulars, irregular buff spotting on nape forming indistinct collar, and irregularly buff spotting on coverts. (See also Table 1 on p.241.) Churring calls similar to those of Eurasian but softer and reminiscent of a distant two-stroke engine, plus a low, soft *chuck-chuck* in flight. Common resident, locally migratory, preferring desert tracts over the more rocky habitats favoured by Eurasian. Absent from most of Baluchistan, northern Punjab and mountainous areas.

4 LARGE-TAILED NIGHTJAR *Caprimulgus macrurus*   33 cm

a b MALE and **c** FEMALE More warmly coloured and strongly patterned than Grey, with longer and broader tail. Has diffuse, pale rufous-brown nuchal collar, well-defined buff edges to scapulars, and prominent buff tips to wing-coverts that usually form pronounced wing-bars. (See also Table 1 on p.241.) Song is a series of loud, resonant calls: *chaunk-chaunk-chaunk*.... Summer breeding visitor in dry subtropical deciduous forest, confined to the Murree Hills eastwards to Kahuta.

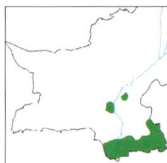

5 INDIAN NIGHTJAR *Caprimulgus asiaticus*   24 cm

a b ADULT Like a small version of Large-tailed. Has boldly streaked crown, rufous-buff nuchal collar, bold black centres and broad buff edges to scapulars, and relatively unmarked central tail feathers. (See also Table 1 on p.241.) Song is a far-carrying *chuk-chuk-chuk-chuk-tukaroo*; also a short, sharp *qwit-qwit* in flight. Locally migratory resident, confined to lower Sind and Dadu District, preferring low hill country or saline flats dotted with tamarisk or thorn scrub.

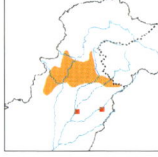

6 SAVANNA NIGHTJAR *Caprimulgus affinis*   23 cm

a b MALE and **c** FEMALE Crown and mantle finely vermiculated, and often appears rather plain except for scapulars, which are edged with rufous buff. Male has largely white outer tail feathers, although this can be difficult to see since the tail is often not fully spread when in flight. (See also Table 1 on p.241.) Song is a strident *dheet*. Common spring migrant, staying to breed throughout northwestern dry foothill areas. Leaves October/November.

1 **ROCK PIGEON** *Columba livia*   33 cm
a b ADULT Grey tail with blackish terminal band, and broad black bars across greater coverts and tertials/secondaries. Back varies from whitish to pale grey (note difference in tail pattern compared with Hill Pigeon). Feral birds vary considerably in coloration and patterning. Abundant throughout Pakistan except in the extreme eastern desert borders, and roosts in uninhabited areas down open wells, on rocky cliffs and on iron girders of bridges. Northern population locally migratory in winter to the Punjab plains. In towns, often mixes with feral pigeons.

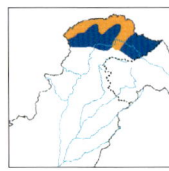
2 **HILL PIGEON** *Columba rupestris*   33 cm
a b ADULT Similar in appearance to Rock Pigeon, but has a white band across the tail that contrasts with the blackish terminal band. Often in large flocks. Resident and confined to higher alpine valleys in extreme northern regions at 2000–5500 m. In winter, flocks are sympatric with Rock Pigeon.

3 **SNOW PIGEON** *Columba leuconota*   34 cm
a b c ADULT Slate-grey head, creamy-white collar and underparts, fawn-brown mantle, and white band across black tail. Keeps in pairs and small parties in summer and in large flocks in winter. Less commensally adapted than Hill Pigeon, and resident in high alpine regions, preferring more mesic alpine slopes than the latter. Migrates in winter down to broader mountain valleys at 1500 m.

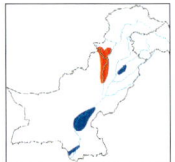
4 **YELLOW-EYED PIGEON** *Columba eversmanni*   30 cm
a b ADULT Smaller than Rock, with narrower and shorter black wing-bars. Yellow orbital skin and iris, yellow tip to bill, brownish cast to upperparts, purplish cast to grey crown and nape, and extensive greyish-white back and upper rump. Also slight differences in tail pattern from Rock: dark terminal band is less clear cut and shows diffuse paler grey subterminal band. Highly gregarious winter visitor to fallow fields adjacent to the Indus river in northern Sind, and less commonly adjacent to the Chenab river in Jhang District. Migrants from Afghanistan breeding grounds pass through Kurram Valley and Kohat. Globally threatened (Vulnerable).

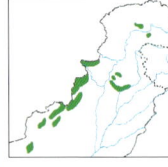
5 **COMMON WOOD PIGEON** *Columba palumbus*   43 cm
a b ADULT White wing-patch and dark tail-band, buff neck-patch and deep vinous underparts. In flight, from below, shows greyish-white band across tail, blackish terminal band, and pale grey undertail-coverts and base of tail. Uncommon resident, mainly sedentary in uncultivated hilly tracts of the Punjab Salt Range, and in hilly tracts along western borders of Baluchistan. Rare in higher valleys of southern Gilgit.

6 **SPECKLED WOOD PIGEON** *Columba hodgsonii*   38 cm
a MALE and **b** FEMALE Speckled underparts, and white spotting on wing-coverts. Male has maroon mantle and maroon on underparts, replaced by grey on female. In flight, from below, shows dark grey undertail-coverts and underside to tail. This frugivorous pigeon is locally migratory and rare in Pakistan, with few scattered records from Indus Kohistan, Hazara District and the Neelum Valley, often around 2000–3000 m.

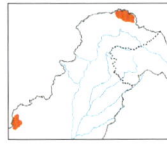

1 **EUROPEAN TURTLE DOVE** *Streptopelia turtur*   33 cm
a b ADULT Told from Oriental by smaller size and slimmer build; broader, paler rufous-buff fringes to scapulars and wing-coverts; more buffish- or brownish-grey rump and uppertail-coverts; and greyish-pink breast, becoming whitish on belly and undertail-coverts. Occasional migrant to Baluchistan in vicinity of Chaman, and to northern Gilgit, mainly in the spring.

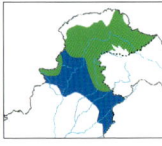

2 **ORIENTAL TURTLE DOVE** *Streptopelia orientalis*   33 cm
a b ADULT Stocky dove with rufous-scaled scapulars and wing-coverts, vinaceous-pink underparts, and black and bluish-grey barring on neck sides. Has dusky-grey underwing. Open forest, especially near cultivation. Widespread resident across northern areas from Chitral to Baltistan, with winter migration to adjacent plains, as far south as Changa Manga Forest Plantation.

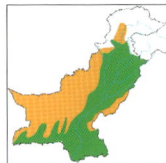

3 **LAUGHING DOVE** *Streptopelia senegalensis*   27 cm
a b ADULT Slim, small dove with fairly long tail. Brownish-pink head and underparts, uniform upperparts, and black stippling on upper breast. Dry cultivation and scrub-covered hills. Abundant resident through plains of Pakistan, extending into eastern desert areas, and summer breeding in hilly tracts of Baluchistan and NWFP.

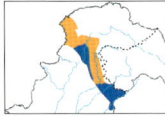

4 **SPOTTED DOVE** *Streptopelia chinensis*   30 cm
a b ADULT Spotted upperparts, and black-and-white chequered patch on neck sides. Mainly summer breeding visitor to scrub-forested slopes in the outer foothills of the Himalayas, from Chitral to Punjab, at 1000–1800 m. A few over-winter in all these areas.

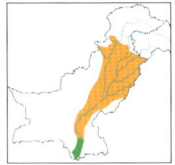

5 **RED COLLARED DOVE** *Streptopelia tranquebarica*   23 cm
a MALE and **b** FEMALE Small, stocky dove, with shorter tail than Eurasian Collared. Male has blue-grey head, pinkish-maroon upperparts and pink underparts. Female similar to Eurasian Collared, but is more compact, with darker buffish-grey underparts, darker fawn-brown upperparts and greyer underwing-coverts. Common summer breeding visitor throughout Indus plains, extending into lower valleys of NWFP. Less common but resident in lower Sind. Prefers riverine forest and tree-lined canals.

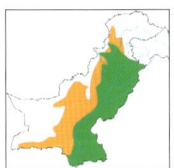

6 **EURASIAN COLLARED DOVE** *Streptopelia decaocto*   32 cm
a b ADULT Sandy brown with black half-collar. Larger and longer tailed than Red Collared. Plumage similar to female Red Collared but with paler upperparts and underparts, and white underwing-coverts. Open dry country with cultivation and groves. Abundant resident, extending further into desert tracts than Laughing Dove. Summer breeder in western parts of Baluchistan, and NWFP north to Chitral.

7 **YELLOW-FOOTED GREEN PIGEON**
Treron phoenicoptera   33 cm
ADULT Large size and greenish coloration. Has broad olive-yellow collar, pale greyish-green upperparts, mauve shoulder-patch, and yellow legs and feet. Mainly winter visitor and locally nomadic, attracted to Pipal and Banyan trees, and, in spring, to irrigated mulberry plantations. Occasional breeding recorded in Punjab. Absent in Sind and west of the Indus.

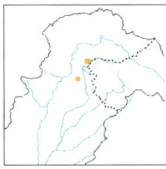

8 **WEDGE-TAILED GREEN PIGEON** *Treron sphenura*   33 cm
a MALE and **b** FEMALE Green upperparts, with long, wedge-shaped tail and yellow edges to wing-coverts and tertials. Male has maroon patch on upperparts, and orange wash to crown and breast. Female has uniform green head. (See Appendix for description of Orange-breasted Green Pigeon.) Mixed broadleaved forest. Rare and probably extirpated in Murree Hills. A small population survives around Malkandi in the Kaghan Valley.

BUSTARDS AND LESSER FLORICAN, PLATE 20

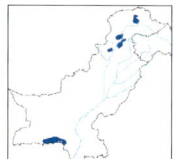

1 **LITTLE BUSTARD** *Tetrax tetrax*   40–45 cm
a b MALE BREEDING, **c** MALE NON-BREEDING and **d e** FEMALE Small, stocky bustard with white panel across secondaries and inner primaries. Both non-breeding male and female are buffish in coloration, with variable streaking and barring (non-breeding male is less heavily marked on upperparts and whiter on underparts compared with female, and shows more white on wing). Breeding male has grey face, and black-and-white pattern on neck and breast. When flushed, has rapid 'winnowing' flight action on stiff, bowed wings. Rare and erratic winter visitor, mainly in open stony or grassy plains, adjacent to hills. Recent records are from Las Bela, Baluchistan, and around Swabi and Jamrud in NWFP.

2 **INDIAN BUSTARD** *Ardeotis nigriceps*   92–122 cm
a b MALE and **c** FEMALE Very large bustard. In all plumages, has greyish or white neck, black crown and crest, uniform brown upperparts, and white-spotted black wing-coverts. Upperwing lacks extensive area of white. Male huge, with black breast-band, and with almost white neck only very finely vermiculated with dark grey. Female smaller; neck appears greyer owing to profuse dark grey vermiculations, and typically lacks black breast-band. Dry grassland with bushes. Depending on good monsoon rains, is an erratic winter visitor to the eastern border regions of the Cholistan and Thar deserts, generally occurring in small groups. Globally threatened (Endangered).

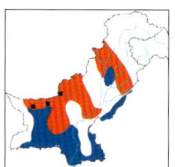

3 **MACQUEEN'S BUSTARD** *Chlamydotis macqueenii*   55–65 cm
a b MALE Medium-sized bustard. In all plumages, shows dark vertical stripe down neck. In flight, extensive white patch on outer primaries. Sexes similar, but female is smaller and lacks whitish panel across greater coverts. Juvenile very similar to female, but lacks black-tipped crest, neck-stripe is finer, and white on wing is washed with buff and is less prominent. (See Appendix for description of vagrant Great Bustard.) Semi-desert with scattered shrubs and sandy grassland. Has been heavily persecuted by Arab hunting parties since the 1960s, but remains a scarce winter visitor to the Thar and Cholistan deserts and across much of southern Baluchistan. In the 1980s, attempted breeding took place in the northern Chagai Desert. Migrant birds enter Pakistan on a broad front through northwest Pakistan. Rare and declining. Globally threatened (Vulnerable).

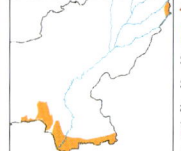

4 **LESSER FLORICAN** *Sypheotides indica*   46–51 cm
a b MALE BREEDING, **c** MALE NON-BREEDING and **d e** FEMALE Small, slim, long-necked bustard (especially compared with Little). Male breeding has spatulate-tipped head plumes, black head, neck and underparts, and white collar across upper mantle; white wing-coverts show as patch on closed wing. Non-breeding male is similar to female but has white wing-coverts. Female and immature are sandy or cinnamon-buff in coloration and cryptically marked; note their different wing pattern, with barred flight feathers, compared with Little and Macqueen's bustards. Occasionally enters southern Sind and Las Bela, as well as the border areas near Kasur, after exceptionally good monsoon rains. Favours grassy plains, avoiding the drier desert tracts preferred by Macqueen's Bustard. Globally threatened (Endangered).

1 **SARUS CRANE** *Grus antigone* 156 cm

a b ADULT and **c** IMMATURE Adult is grey, with bare red head and upper neck, and bare ashy-green crown. In flight, black primaries contrast with rest of wing. Immature has rusty-buff feathering to head and neck, and upperparts are marked with brown; older immatures are similar to adult but have dull red head and upper neck, and lack greenish crown of adult. Wetlands and irrigated agricultural fields. One pair of this sedentary species was recently found breeding in the Badin District on the Rann of Kutch border, with occasional records of single birds on northeastern borders of Punjab and southern Sind. Globally threatened (Vulnerable).

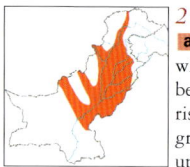

2 **DEMOISELLE CRANE** *Grus virgo* 90–100 cm

a b ADULT and **c** IMMATURE Small crane. Adult has black head and neck with white tuft behind eye, and grey crown; black neck feathers extend as a point beyond breast, and elongated tertials project as shallow arc beyond body, giving rise to distinctive shape. Immature similar to adult, but head and neck are dark grey, tuft behind eye is grey and less prominent, and it has grey-brown cast to upperparts. In flight, Demoiselle is best separated at a distance from Common by black breast. Cultivation and large rivers. Once abundant in winter, now only a spring and autumn passage migrant through Pakistan, with most records from Punjab, especially flocks resting in the desert borders and entering or leaving along the western boundaries of Baluchistan and NWFP. Highly prized as a pet and ruthlessly hunted for as such by Wazir tribesmen.

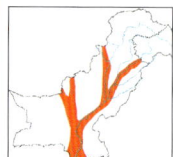

3 **COMMON CRANE** *Grus grus* 110–120 cm

a b ADULT and **c** IMMATURE Adult has mainly black head and fore-neck, with white stripe behind eye extending down side of neck. Immature has brown markings on upperparts, with buff or grey head and neck. Adult head pattern apparent on some by first winter, and as adult by second winter. Cultivation, large rivers and marshes. Scarce passage migrant, with numbers declining due to heavy hunting pressure in Kohat and Bannu districts, where it is mainly captured alive using decoys. In the early 1980s, surveys revealed that about 750 Common and the same number of Demoiselle cranes were captured annually, with over 5000 birds held in villages in the NWFP.

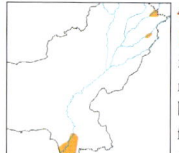

4 **WATERCOCK** *Gallicrex cinerea* M 43 cm, F 36 cm

a MALE BREEDING, **b** JUVENILE MALE and **c** FEMALE Male is mainly greyish black, with yellow-tipped red bill and red shield and horn. Non-breeding male and female have buff underparts with fine barring, and buff fringes to dark brown upperparts. Juvenile has uniform rufous-buff underparts and rufous-buff fringes to upperparts. Male is much larger than female. Males call incessantly, uttering extraordinary pumping calls, *qhumb-qhumb qhumb*. Marshes and flooded fields. Uncommon summer breeding visitor, entering northeast Punjab along the Chenab and Ravi riverine areas, but more plentiful in lower Sind, spreading through rice-growing tracts.

5 **PURPLE SWAMPHEN** *Porphyrio porphyrio* 45–50 cm

a ADULT and **b** JUVENILE Large size, purplish-blue coloration, and huge red bill and red frontal shield. Juvenile greyer, with duller bill. Throughout year very vocal, with peculiar squawking, wailing and cackling, eliciting a synchronised response. Reed-beds and marshes. Year-round resident throughout wetlands of Sind and Punjab. Locally migratory post-monsoon to new feeding areas. Haleji Reservoir, Thatta District, has a permanent population of about 1200.

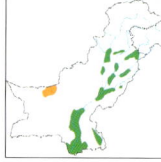

RAILS AND CRAKES, PLATE 22

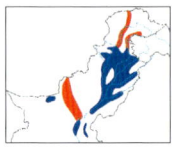

1 WATER RAIL *Rallus aquaticus*   23–28 cm
a ADULT and **b** JUVENILE Slightly down-curved bill with red at base. Legs pinkish. Adult has streaked upperparts, greyish underparts, and barring on flanks. Juvenile has buff coloration to underparts. Crepuscular choruses of squealing, even in winter. Marshes and wet fields. Widespread winter visitor, particularly in northern Punjab. Comparatively rare in Sind.

2 BROWN CRAKE *Amaurornis akool*   28 cm
ADULT Olive-brown upperparts, grey underparts, and olive-brown undertail-coverts; lacks barring. Has greenish bill and pinkish legs. Calls are a rapid whistle. Marshes. Rare resident in Rawalpindi District.

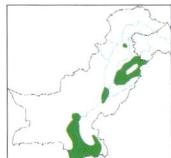

3 WHITE-BREASTED WATERHEN
Amaurornis phoenicurus   32 cm
a ADULT and **b** JUVENILE Adult has grey upperparts, and white face, foreneck and breast; undertail-coverts rufous cinnamon. Juvenile duller. Indulges in noisy crepuscular choruses, *kurrwarh-kurrwah*, breaking into *kwokkwok-krrr-oowok-oowok*. Marshes and thick cover close to pools, lakes and ditches. Uncommon resident; southern Sind and northern Punjab.

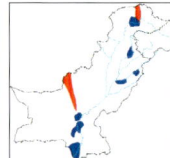

4 LITTLE CRAKE *Porzana parva*   20–23 cm
a MALE, **b** FEMALE and **c** JUVENILE From Baillon's Crake by longer wings (primaries extend noticeably beyond tertials at rest), less extensive barring on flanks, and pronounced pale edges to scapulars and tertials (all plumages). Adult also has red at base of bill. Male has grey underparts. Female and juvenile have buff underparts (but less barring than on Baillon's). Scarce winter visitor favouring dense reed cover. Northern areas, down to Larkana District.

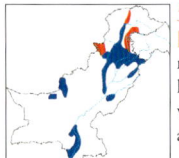

5 BAILLON'S CRAKE *Porzana pusilla*   17–19 cm
a ADULT and **b** JUVENILE Adult has rufous-brown upperparts, extensively marked with white. Flanks are barred. Bill and legs are green. Juvenile is similar but has buff underparts. Marshes, reedy lake edges and wet fields. Uncommon winter visitor, with widely scattered records from lower Sind and northern Punjab. Spring and autumn passage noted in Chaman, Baluchistan, and Indus and Jelum valleys.

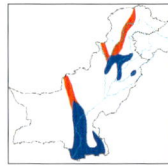

6 SPOTTED CRAKE *Porzana porzana*   22–24 cm
a ADULT and **b** JUVENILE Profuse white spotting on head, neck and breast. Stout bill, barred flanks, and unmarked buff undertail-coverts. Adult has yellowish bill with red at base, and grey head and breast. Juvenile has buffish-brown head and breast, and bill is brown. (See Appendix for description of Corn Crake.) Marshes and lakes. Scarce winter visitor, mainly to northwestern Punjab and Sind from Dadu District south to Tharparkar.

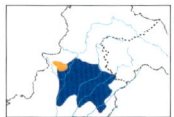

7 RUDDY-BREASTED CRAKE *Porzana fusca*   22 cm
a ADULT and **b** JUVENILE Red legs, chestnut underparts, and black-and-white barring on undertail-coverts. Juvenile is dark olive-brown, with white-barred undertail-coverts and fine greyish-white barring on underparts. Marshes and wet paddyfields. Uncommon in Punjab Salt Range and plains of western NWFP. Occasionally breeds Kurram Valley.

8 COMMON MOORHEN *Gallinula chloropus*   32–35 cm
a ADULT and **b** JUVENILE White undertail-coverts and line along flanks. Adult has red bill with yellow tip and red shield. Juvenile has dull green bill and is mainly brown. Abundant resident, inhabiting all types of wetlands with reed cover, avoiding only large, open water areas. Absent from dry mountainous areas.

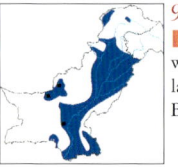

9 COMMON COOT *Fulica atra*   36–38 cm
a ADULT and **b** JUVENILE Blackish, with white bill and shield. Abundant winter visitor throughout Indus basin, favouring large, open waterbodies, with large counts from Khinjir Lake and Zangi Nawar Lake. Occasional breeding in Baluchistan and Larkana District.

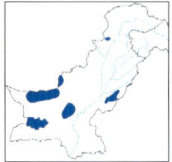

1 **PIN-TAILED SANDGROUSE** *Pterocles alchata* 31–39 cm
a b MALE BREEDING and **c d** FEMALE Pin-tailed. White belly, with 2 (male) or 3 (female) narrow black bands across neck and breast. Largely white underwing and pale grey upperside to primaries. Male in breeding plumage has greenish upperparts with yellowish spotting; buff and barred with black in non-breeding plumage. Calls are a guttural *quattarr*. Desert and semi-desert. Least plentiful of sandgrouse in Pakistan. Confined mainly to Chagai and Sibi deserts as winter visitor. Occasional breeding recorded on western borders.

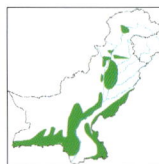

2 **CHESTNUT-BELLIED SANDGROUSE**
Pterocles exustus 31–33 cm
a b MALE and **c d** FEMALE Pin-tailed, with dark underwing, blackish-chestnut belly and black breast line. Female has buff banding across upperwing-coverts and lacks black gorget across throat, which are useful distinctions at rest from Black-bellied. Calls are a liquid *kwit-kwit-kwiturrohkwiturroh*. Desert and sparse thorn scrub. Common and widespread resident, avoiding irrigated tracts and mountainous regions. Breeds in the Punjab Salt Range, Cholistan and Thar deserts, and dry valleys in southern Baluchistan.

3 **SPOTTED SANDGROUSE** *Pterocles senegallus* 30–35 cm
a b MALE and **c** FEMALE Pin-tailed. Rather pale upperwing with dark trailing edge, and whitish belly with black line down centre. Female is spotted on upperparts and breast. Calls are louder and slower than those of Chestnut-bellied, *wittoo-wittoo*. Sandy desert and arid foothills. Winter visitor to drier desert and stony plains, southern Baluchistan, and Cholistan and Thar deserts. Absent from northern Baluchistan and Punjab. Occasional breeding in foothills on west bank of Indus.

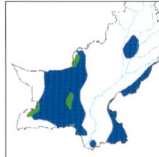

4 **BLACK-BELLIED SANDGROUSE**
Pterocles orientalis 33–35 cm
a b MALE and **c** FEMALE Stocky and short-tailed. Both sexes have black belly, and white underwing-coverts contrast with black flight feathers. Male has black and chestnut throat and grey neck and breast. Female is very heavily marked. Calls are a gruff *katarr-katarr*. Common winter visitor throughout southern Baluchistan, and Thal and Cholistan deserts, but curiously rare in Thar Desert. Occasional breeding in southern Kalat and Chagai. Prefers scrub desert and abandoned cultivation.

5 **CROWNED SANDGROUSE** *Pterocles coronatus* 27–29 cm
a b MALE and **c** FEMALE Small, compact and short-tailed. Blackish flight feathers contrast with coverts on both surfaces of wing. Male has bold buff spotting on scapulars and upperwing-coverts, and black-and-white markings on head. Female uniformly barred and spotted, including on belly, and has orange-buff throat. Calls are a high-pitched *quittoo-quitoo* and *quit-quit-quit-quidu-ke-quidu*. Desert. Uncommon, largely resident in most barren arid tracts of Mekran and Kalat in southern Baluchistan, east to Sind Kohistan. Not recorded east of the Indus.

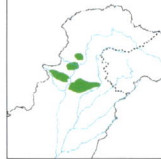

6 **PAINTED SANDGROUSE** *Pterocles indicus* 28 cm
a b MALE and **c** FEMALE Small, stocky and short-tailed. Heavily barred. Underwing dark grey. Male has chestnut, buff and black bands across breast, and unbarred orange-buff neck and inner wing-coverts. Female heavily barred all over, with yellowish face and throat. Uncommon resident in low, arid hilly tracts in the Punjab Salt Range west to Kohat. Less gregarious than other sandgrouse, and only comes to drink after dark.

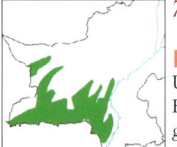

7 **LICHTENSTEIN'S SANDGROUSE**
Pterocles lichtensteinii 24–26 cm
a b MALE and **c** FEMALE Small, stocky and short-tailed. Heavily barred. Underwing dark grey. Male has buff and black bands across breast and barred neck. Female heavily barred all over, including on face and throat. Calls are a lower pitched, grating *gwittoogwittoo*. Resident and commoner than Painted, preferring low rocky desert hills and valleys throughout Mekran and Sind Kohistan. Absent east of the Indus. Like the Painted, it drinks only after dark.

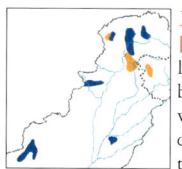

1 **EURASIAN WOODCOCK** *Scolopax rusticola*   33–35 cm
a b ADULT Bulky, with broad, rounded wings. Head banded black and buff; lacks sharply defined mantle and scapular stripes. Crepuscular and nocturnal. In breeding season, male has characteristic roding display-flight at dawn and dusk, when it flies above the treetops with slow, deliberate wingbeats, uttering *chiwich* call. Rare breeding resident in moister temperate Himalayan forest from Chitral to Neelum Valley, migrating altitudinally in winter, with marked passage through Uruk Valley, Baluchistan and Kurram Valley, NWFP.

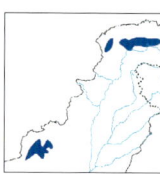

2 **SOLITARY SNIPE** *Gallinago solitaria*   29–31 cm
a b ADULT Large, dull-coloured snipe with long bill. Colder-coloured and less boldly marked than Common Snipe, with less striking head pattern, white spotting on ginger-brown breast, and fine white mantle and scapular stripes. Has fine white trailing edge to wing, and underwing is heavily barred. When flushed, does not rise steeply, flying more heavily and slowly than Common, giving a harsh *kensh* call and soon settling again. In aerial display, utters a deep *chok-achok* call, combined with a mechanical bleating produced by outer tail feathers. Marshy edges and beds of mountain streams. Rare resident of high alpine regions of Karakorams, and Kalpiphat, Zarghun ranges in Baluchistan. Winters in adjacent river valleys.

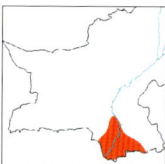

3 **PINTAIL SNIPE** *Gallinago stenura*   25–27 cm
a b c ADULT More rounded wings than Common, and slower and more direct flight. Lacks white trailing edge to secondaries, and has densely barred underwing-coverts and pale (buff-scaled) upperwing-covert panel (more pronounced than shown). Feet project noticeably beyond tail in flight. Flight call is a rasping *tetch*. Marshes and wet paddy-fields. Spring and autumn passage migrant; recorded only in lower Sind, where it is fairly common in September and March.

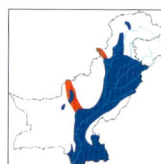

4 **COMMON SNIPE** *Gallinago gallinago*   25–27 cm
a b c ADULT Compared with Pintail, wings are more pointed and it has faster and more erratic flight; shows prominent white trailing edge to wing and white banding on underwing-coverts. Flight call is a grating *scaaap*, higher pitched and more anxious than that of Pintail. Marshes and wet paddy stubbles. Common winter visitor throughout the Indus plains, arriving end of August and leaving March/April. Strong passage across central Baluchistan from Chaman and Kurram Valley, NWFP.

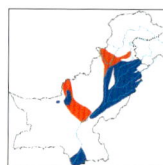

5 **JACK SNIPE** *Lymnocryptes minimus*   17–19 cm
a b c ADULT Small, with short bill. Flight weaker and slower than that of Common, and has rounded wing-tips. Has divided supercilium but lacks pale crown-stripe. Mantle and scapular stripes very prominent. If flushed, flies off silently and without zigzagging flight. Marshes and wet paddy stubbles. Uncommon winter visitor to Punjab and lower Sind, with marked passage through Kurram Valley, NWFP and central Baluchistan.

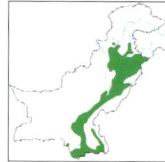

6 **GREATER PAINTED-SNIPE** *Rostratula benghalensis*   25 cm
a ADULT MALE, **b** ADULT FEMALE and **c** JUVENILE Rail-like wader, with broad, rounded wings and longish, down-curved bill. White or buff 'spectacles' and 'braces'. Adult female has maroon head and neck, and dark greenish wing-coverts. Adult male and juvenile are duller, and have buff spotting on wing-coverts. When flushed, rises heavily with legs trailing. Polyandrous females call at night, *khoonk-khoonk* up to 40 times. Marshes, vegetated pools and stream banks; tolerant of brackish salt marshes, but prefers major rice-growing tracts. Resident but locally nomadic throughout the Indus and its tributaries.

GODWITS, CURLEWS AND *TRINGA* SANDPIPERS, PLATE 25

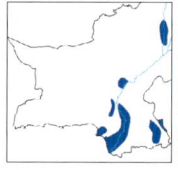

1 **BLACK-TAILED GODWIT** *Limosa limosa* 36–44 cm
a MALE BREEDING, **b** NON-BREEDING, **c** JUVENILE and **d** FLIGHT White wing-bars and white tail-base with black tail-band. In breeding plumage, male has rufous-orange neck and breast, with blackish barring on underparts and white belly; breeding female is duller. In non-breeding plumage, adult is uniform grey on neck, upperparts and breast. Juvenile has cinnamon underparts and cinnamon fringes to dark-centred upperparts. Highly gregarious winter visitor, confined to larger inland wetlands, mostly in Sind, with small non-breeding groups summering in paddy-fields.

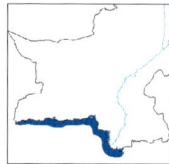

2 **BAR-TAILED GODWIT** *Limosa lapponica* 37–41 cm
a MALE BREEDING, **b** NON-BREEDING, **c** JUVENILE and **d** FLIGHT Lacks wing-bar, and has barred tail and white V on back. Breeding male has chestnut-red underparts. Breeding female has cinnamon underparts, although many as non-breeding. Non-breeding adult has dark streaking on breast and streaked appearance to upperparts. Juvenile similar, but with buff wash to underparts and buff edges to mantle/scapulars. Winter visitor to estuaries and lagoons of Arabian Sea coast.

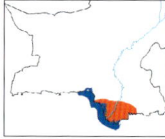

3 **WHIMBREL** *Numenius phaeopus* 40–46 cm
a **b** ADULT Smaller than Eurasian Curlew, with shorter bill. Distinctive head pattern, with whitish supercilium and crown-stripe, dark eye-stripe and dark sides of crown. Flight call distinctive, *he-he-he-he-he-he-he*. Paddy-fields, mudflats and sea coasts. Passage migrant, particularly through lower Sind, with birds remaining along the Karachi sea coast and, particularly, in Indus delta mangroves.

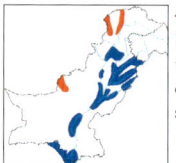

4 **EURASIAN CURLEW** *Numenius arquata* 50–60 cm
a **b** ADULT Large size and long, curved bill. Rather plain head. Has distinctive mournful *cur-lew* call. Winter visitor, uncommon on margins of inland lakes, common along Arabian Sea coast and mangroves. Over 5000 counted during survey of Indus delta.

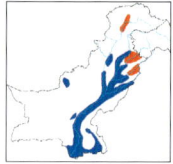

5 **SPOTTED REDSHANK** *Tringa erythropus* 29–32 cm
a ADULT BREEDING, **b** **c** ADULT NON-BREEDING and **d** JUVENILE Red at base of bill, and red legs. Longer bill and legs than Common Redshank, and upper-wing uniform. Non-breeding plumage is paler grey above and whiter below than Common. Underparts mainly black in breeding plumage. Juvenile has grey barring on underparts. Has distinctive *tu-ick* flight call. Uncommon winter visitor to wetlands of Indus basin. Tolerant of brackish pools.

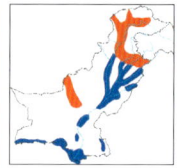

6 **COMMON REDSHANK** *Tringa totanus* 27–29 cm
a ADULT BREEDING, **b** **c** ADULT NON-BREEDING and **d** JUVENILE Legs and bill base orange-red. Shorter bill and legs than Spotted Redshank, and with broad white trailing edge to wing. Non-breeding plumage is grey-brown above, with grey breast. Neck and underparts heavily streaked in breeding plumage. Juvenile has brown upperparts with buff spotting. Call is an anxious *teu-hu-hu*. Common winter visitor, mainly along the sea coast, especially on mudbanks.

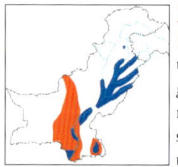

7 **MARSH SANDPIPER** *Tringa stagnatilis* 22–25 cm
a ADULT BREEDING and **b** **c** ADULT NON-BREEDING Smaller and daintier than Common Greenshank, with proportionately longer legs and finer bill. Legs greenish or yellowish. In non-breeding plumage, upperparts are grey and fore-neck and underparts are white. In breeding plumage, fore-neck and breast are streaked and upperparts are blotched and barred. Juvenile has dark-streaked upperparts with buff fringes. Has an abrupt, dull *yup* flight call. Uncommon winter migrant with strong passage through central Baluchistan. Prefers freshwater wetlands.

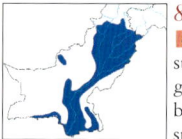

8 **COMMON GREENSHANK** *Tringa nebularia* 30–34 cm
a ADULT BREEDING and **b** **c** ADULT NON-BREEDING Stocky, with long, stout bill and long, stout greenish legs. In non-breeding plumage, upperparts are grey and fore-neck and underparts are white. In breeding plumage, fore-neck and breast are streaked and upperparts are untidily streaked. Juvenile has dark-streaked upperparts with fine buff or whitish fringes. Call is a loud, ringing *tu-tu-tu*. Abundant winter visitor to wetlands throughout Pakistan. A few non-breeders over-summer.

MISCELLANEOUS WADERS, PLATE 26

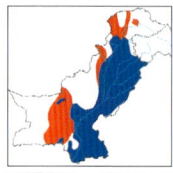

1 **GREEN SANDPIPER** *Tringa ochropus*   21–24 cm
a b c ADULT NON-BREEDING Greenish legs. White rump; dark upperwing and underwing. Compared with Wood, has indistinct (or non-existent) supercilium behind eye and darker upperparts. *Tluee-tueet* flight call. Common winter visitor all over Pakistan except dry mountainous areas. Able to exploit isolated small wetlands unattractive to other waders. A few non-breeders over-summer.

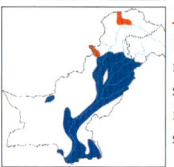

2 **WOOD SANDPIPER** *Tringa glareola*   18–21 cm
a b c ADULT NON-BREEDING and **d** JUVENILE Yellowish legs. White rump, and upperwing lacks wing-bar. Compared with Green, shows prominent supercilium, more heavily spotted upperparts, and paler underwing. Flight call is a soft *chiff-if-if*. Common winter visitor throughout Indus basin, preferring shallow flooded areas and rice stubbles, and avoiding sea coast, except on passage.

3 **TEREK SANDPIPER** *Xenus cinereus*   22–25 cm
a b ADULT BREEDING, **c** ADULT NON-BREEDING and **d** JUVENILE Longish, upturned bill and short yellowish legs. In flight, shows prominent white trailing edge to secondaries and grey rump and tail. Adult breeding has blackish scapular lines. Flight call is a pleasant *hu-hu-hu*. Winter visitor common along Arabian Sea coast, especially in mangroves. Scattered records inland on passage.

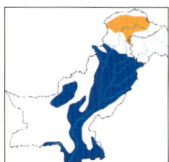

4 **COMMON SANDPIPER** *Actitis hypoleucos*   19–21 cm
a b ADULT and **c** JUVENILE Horizontal stance and constant bobbing action. In flight, rapid shallow wingbeats are interspersed with short glides. Juvenile has buff fringes to upperparts. Flight call is an anxious *wee-wee-wee*. Not abundant but widespread winter visitor inland, preferring major canals, village tanks and, on the coast, muddy sheltered creeks. Breeds on all northern mountain rivers from Chitral east to Baltistan.

5 **GREAT KNOT** *Calidris tenuirostris*   26–28 cm
a ADULT BREEDING, **b c** ADULT NON-BREEDING and **d** JUVENILE Larger than Red Knot (see Appendix), and often with slightly down-curved bill. Adult breeding is heavily marked with black on breast and flanks, with chestnut patterning to scapulars. Adult non-breeding is typically more heavily streaked on upperparts and breast than Red Knot, and juvenile is more strongly patterned than that species, with blackish centres to mantle and scapulars, and more heavily marked breast and flanks. Rare winter visitor along intertidal zone of Arabian Sea. Usually in flocks, staying until late May.

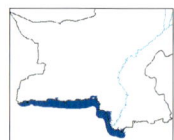

6 **SANDERLING** *Calidris alba*   20 cm
a ADULT BREEDING, **b c** ADULT NON-BREEDING and **d** JUVENILE Stocky, with short bill. Very broad, white wing-bar. Adult breeding usually shows some rufous on sides of head, breast and upperparts. Non-breeding is pale grey above and very white below. Juvenile is chequered black and white above. Winter visitor to sandy beaches all along Arabian Sea coast.

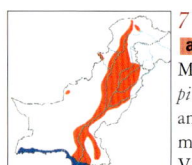

7 **LITTLE STINT** *Calidris minuta*   13–15 cm
a ADULT BREEDING, **b c** ADULT NON-BREEDING and **d e** JUVENILE More rotund than Temminck's, with dark legs. Grey sides to tail; has weak *pi-pi-pi* flight call. Adult breeding has pale mantle V, rufous wash to face, neck sides and breast, and rufous fringes to upperpart feathers. Non-breeding has untidy, mottled/streaked appearance, with grey breast sides. Juvenile has whitish mantle V, greyish nape, and rufous fringes to upperparts. Mudflats and muddy wetlands; more common inland on passage. Highly gregarious and very abundant winter visitor all along Arabian Sea coast, migrating through northern mountain valleys.

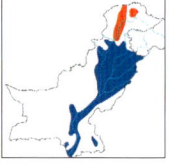

8 **TEMMINCK'S STINT** *Calidris temminckii*   13–15 cm
a ADULT BREEDING, **b** ADULT NON-BREEDING and **c d** JUVENILE More elongated and horizontal than Little. White sides to tail in flight; flight call is a purring trill. Legs yellowish. In all plumages, lacks mantle V and is usually rather uniform, with complete breast-band and indistinct supercilium. Abundant winter visitor, preferring inland freshwater marshes, pools and even freshly irrigated fields. Much less common in Sind.

MISCELLANEOUS WADERS, PLATE 27

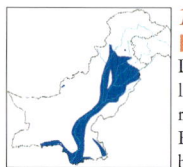

1 **DUNLIN** *Calidris alpina* 16–22 cm
a b ADULT BREEDING, **c d** ADULT NON-BREEDING and **e** JUVENILE
Dark centre to rump. Adult breeding has black belly. Adult non-breeding has less distinct supercilium than Curlew Sandpiper. Juvenile has streaked belly, rufous fringes to mantle, and buff mantle V. (See Appendix for comparison with Red Knot.) Flight call is a slurred *screet*. Abundant winter visitor to mudflats and beaches along Arabian Sea coast, but also favouring sandbars and oxbow lakes along Indus and its tributaries.

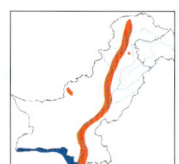

2 **CURLEW SANDPIPER** *Calidris ferruginea* 18–23 cm
a b ADULT BREEDING, **c d** ADULT NON-BREEDING and **e** JUVENILE
White rump. More elegant than Dunlin, with longer and more down-curved bill. Adult breeding has chestnut-red underparts. Adult non-breeding has more distinct supercilium than Dunlin. Juvenile has strong supercilium, buff wash to breast, and buff fringes to upperparts. (See Appendix for comparison with Red Knot.) Flight call is a purring *prrriit*. Common winter visitor to mudflats and beaches all along Arabian Sea coast, with double passage from end of July, returning mid-May through Indus and Shyok valleys.

3 **BROAD-BILLED SANDPIPER** *Limicola falcinellus* 16–18 cm
a b ADULT BREEDING, **c d** ADULT NON-BREEDING and **e** JUVENILE
Distinctive shape: stockier than Dunlin, with legs set well back and downward-kinked bill. In all plumages, has more prominent supercilium than Dunlin, with 'split' before eye, and contrasting with dark eye-stripe. Adult breeding has bold streaking on neck and breast contrasting with white belly. Non-breeding has dark patch at bend of wing (sometimes obscured by breast feathers) and strong streaking on breast; dark inner wing-coverts show as dark leading edge to wing in flight. Juvenile has buff mantle/scapular lines and streaked breast. Flight call is a buzzing *chrrreet*. Uncommon winter migrant, preferring muddy embankments and creeks along sea coast, with influx of spring migrants from India. About 400 counted on Gizri Creek, Indus delta, in April.

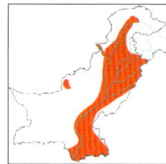

4 **RUFF** *Philomachus pugnax* M 26–32 cm, F 20–25 cm
a b c MALE BREEDING, **d e** NON-BREEDING and **f** JUVENILE Males larger than females. Non-breeding and juvenile have neatly fringed upperparts. Breeding male has variable ruff. Marshes, wet fields, banks of rivers and lakes. Mainly passage migrant, wintering predominantly in Sri Lanka. Prefers inland wetlands; rarely seen on sea coast.

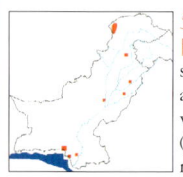

5 **RED-NECKED PHALAROPE** *Phalaropus lobatus* 18–19 cm
a ADULT BREEDING, **b c** ADULT NON-BREEDING and **d** JUVENILE Typically seen swimming. Delicately built with fine bill. Adult breeding has white throat and red stripe down side of grey neck. Adult non-breeding has grey upperparts with white edges to mantle and scapular feathers, forming fairly distinct lines (poorly depicted in plate). Juvenile has dark grey upperparts with orange-buff mantle and scapular lines. (See Appendix for comparison with Red Phalarope.) Abundant winter visitor, occurring in flocks 32–48 km offshore, all along Arabian Sea coast. Prior to spring passage, occasionally sighted on inland wetlands.

6 **RUDDY TURNSTONE** *Arenaria interpres* 23 cm
a ADULT BREEDING, **b c** ADULT NON-BREEDING and **d** JUVENILE Short bill and orange legs. In flight, shows white stripes on wings and back, and black tail. In breeding plumage, has complex black-and-white neck and breast pattern, and much chestnut-red on upperparts; duller and less strikingly patterned in non-breeding plumage. Juvenile has buff fringes to upperparts, and blackish breast. Uncommon winter visitor, favouring Pakistan's few rocky shores, and also foraging on intertidal mudbanks.

JACANAS AND MISCELLANEOUS WADERS, PLATE 28

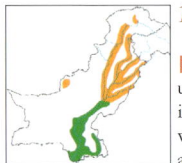

1 PHEASANT-TAILED JACANA
Hydrophasianus chirurgus   31 cm

a b ADULT BREEDING and **c** ADULT NON-BREEDING Extensive white on upperwing, and white underwing. Yellowish patch on sides of neck. Adult breeding has brown underparts and long tail. Adult non-breeding and juvenile have white underparts, with dark line down side of neck and dark breast-band (which are too distinct in plate). Monsoon breeding calls are a cat-like *meyu-meyu-meyu* alternating with *mee-ooph-ooph-ooph-ooph*. Marshes, lakes and ponds with floating vegetation. Common resident in Sind. Summer breeder in Punjab, and main valleys in northern areas.

2 BRONZE-WINGED JACANA *Metopidius indicus*   28–31 cm

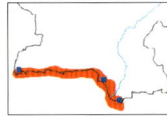

a b ADULT and **c** IMMATURE Dark upperwing and underwing. Adult has white supercilium, bronze-green upperparts and blackish underparts. Juvenile has orange-buff wash on breast, short white supercilium and yellowish bill. Marshes, pools and lakes with floating vegetation. Erratic occurrence in southern Sind only, subject to failing monsoon rains, and sedentary up to 2 years after arrival.

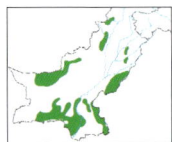

3 CRAB-PLOVER *Dromas ardeola*   38–41 cm

a b ADULT and **c** JUVENILE Black-and-white plumage, with stout black bill and very long blue-grey legs. Juvenile is like washed-out version of adult. Intertidal mudflats, coral reefs and coastal rocks. Uncommon erratic visitor to sea coast, usually in small flocks, with main wintering concentrations in Rann of Kutch, India.

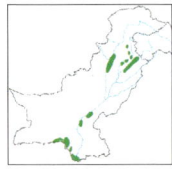

4 EURASIAN THICK-KNEE *Burhinus oedicnemus*   40–44 cm

a b ADULT Sandy brown and streaked. Short yellow-and-black bill, striking yellow eye and long yellow legs. Call is a loud *cur-lee*. Mainly active at dusk and during night, and spends day sitting in shade. A scarce but nomadic resident, favouring main desert tracts and broad stony valleys from Mekran coast up to NWFP.

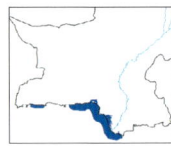

5 GREAT THICK-KNEE *Esacus recurvirostris*   49–54 cm

a b ADULT Upturned black-and-yellow bill, white forehead and 'spectacles', and dark mask. Dark bar across coverts is very prominent at rest. In flight, shows grey mid-wing panel and white patches in primaries. Mainly active at dusk and during the night, but often observable during the day, especially when disturbed. Scarce resident from Arabian Sea coast north to Himalayan foothills. Favours salt marshes or grassy and stony banks alongside estuaries and rivers.

6 EURASIAN OYSTERCATCHER
Haematopus ostralegus   40–46 cm

a b ADULT BREEDING and **c** ADULT NON-BREEDING Black and white, with broad white wing-bar. Bill reddish. White collar in non-breeding plumage. Sandy and rocky coasts, and coral reefs. Abundant sea coast winter visitor, with a few non-breeders staying over the summer.

7 IBISBILL *Ibidorhyncha struthersii*   38–41 cm

a b ADULT and **c** JUVENILE Adult has black face, down-curved dark red bill, and black and white breast-bands. Juvenile has brownish upperparts with buff fringes, faint breast-band, and dull legs and bill. Mountain streams and rivers with shingle beds. Rare resident, with tiny breeding population near Chutran village, Shigar Valley, where hot springs prevent the river from freezing over.

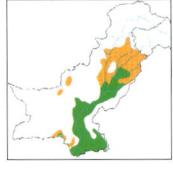

8 BLACK-WINGED STILT *Himantopus himantopus*   35–40 cm

a b ADULT and **c** IMMATURE Black and white, slender appearance, long pinkish legs, and fine straight bill. Upperwing black and legs extend a long way behind tail in flight. Juvenile has browner upperparts with buff fringes. Very common widespread resident. Locally migratory according to feeding conditions, preferring shallow brackish waters and avoiding marine areas. Up to 1600 around Ghauspur Lake in winter.

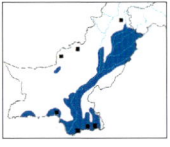

9 PIED AVOCET *Recurvirostra avosetta*   42–45 cm

a b ADULT Upward kink to black bill. Distinctive black-and-white patterning. Juvenile has brown and buff mottling on mantle and scapulars. Marshes and banks of rivers and lakes. Uncommon but erratic visitor throughout Indus basin and along Arabian Sea coast. Breeding recorded on Baluchistan coastal lagoons and Band Khushdil Khan Lake.

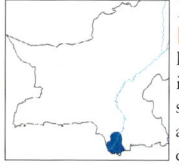

1 PACIFIC GOLDEN PLOVER *Pluvialis fulva*   23–26 cm

a b ADULT BREEDING and **c d** ADULT NON-BREEDING A slim-bodied, long-necked and long-legged plover. In all plumages, has golden-yellow markings on upperparts, and dusky grey underwing-coverts and axillaries. In flight, shows narrower white wing-bar and dark rump. In non-breeding plumage, usually shows prominent pale supercilium and dark patch at rear of ear-coverts (not depicted well in plate). Black underparts with white border are striking in breeding plumage. (See Appendix for comparison with European Golden Plover.) Call is a plaintive *tu-weep*. Mudbanks of wetlands, ploughed fields and grassland. Scarce winter visitor, mainly in lower Sind.

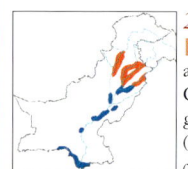

2 GREY PLOVER *Pluvialis squatarola*   27–30 cm

a ADULT BREEDING and **b c d** ADULT NON-BREEDING White underwing and black axillaries. Stockier, with stouter bill and shorter legs, than Pacific Golden. Whitish rump and prominent white wing-bar. Has extensive white spangling to upperparts in breeding plumage; upperparts mainly grey in non-breeding (in all plumages, lacks golden spangling of Pacific Golden). Call is a mournful *chee-woo-ee*. Sandy shores, mudflats and tidal creeks. Common winter visitor, most abundant in Sind and Mekran tidal creeks, with scattered individuals on margins of inland lakes.

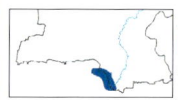

3 COMMON RINGED PLOVER *Charadrius hiaticula*   18–20 cm

a ADULT BREEDING and **b c** ADULT NON-BREEDING Prominent breast-band and white hind-collar. Prominent wing-bar in flight. Adult breeding has orange legs and bill-base (duller in non-breeding; more olive-yellow in juvenile). Non-breeding and juvenile have prominent whitish supercilium and forehead compared with Little Ringed. Flight call is a soft *too-li*, or *too weep* when alarmed. Uncommon winter visitor to Arabian Sea coast.

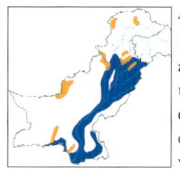

4 LITTLE RINGED PLOVER *Charadrius dubius*   14–17 cm

a ADULT BREEDING and **b c** ADULT NON-BREEDING Small size, elongated and small-headed appearance, and uniform upperwing with only a narrow indistinct wing-bar. Legs yellowish or pinkish. Adult breeding has striking yellow eye-ring. Adult non-breeding and juvenile have less distinct head pattern. Flight call is a clear *peeu*. Common winter visitor to wetlands throughout the Indus basin, with a large resident population breeding sparsely along stony river beds throughout the Indus basin, Baluchistan and NWFP.

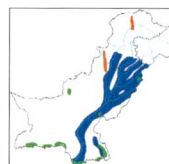

5 KENTISH PLOVER *Charadrius alexandrinus*   15–17 cm

a MALE BREEDING and **b c** ADULT NON-BREEDING Small size and stocky appearance. White hind-collar and usually small, well-defined patches on sides of breast. Male has rufous cap. Flight call is a soft *pi…pi…pi*, or a rattling trill. Common resident, breeding sparsely on suitable desiccating muddy shores of Baluchistan lakes, and commonly in sand dunes along the Mekran coast. Winters throughout the Indus basin.

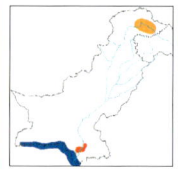

6 LESSER SAND PLOVER *Charadrius mongolus*   19–21 cm

a MALE BREEDING, **b** FEMALE BREEDING and **c d** ADULT NON-BREEDING Larger and longer-legged than Kentish, lacking white hind-collar. Very difficult to distinguish from Greater Sand Plover, although it is smaller and has stouter bill (with blunt tip), and shorter dark grey or dark greenish legs. In flight, feet do not usually extend beyond tail and white wing-bar is narrower across primaries. Breeding male typically shows full black mask and forehead and more extensive rufous on breast compared with Greater Sand (although variation exists in these characters). Flight call is a hard *chitik* or *chi-chi-chi*. Before spring migration, birds in full breeding plumage give harsh grating songs, *trit-it-it-it-tirkhweeoo*. Abundant winter visitor along Arabian Sea coast, frequenting mudflats, sandy beaches, mangroves and rocky shores, with occasional breeding on northern boundaries of Deosai Plateau.

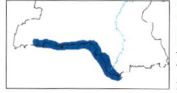

7 GREATER SAND PLOVER *Charadrius leschenaultii*   22–25 cm

a MALE BREEDING, **b** FEMALE BREEDING and **c d** ADULT NON-BREEDING Larger and lankier than Lesser Sand, with longer and larger bill, usually with pronounced gonys and more pointed tip. Longer legs are paler, with distinct yellowish or greenish tinge. In flight, feet project beyond tail and has broader white wing-bar across primaries. Assumes breeding plumage in late April before migration and displays with high twittering trills. Flight call is a trilling *prrrirt*, softer than that of Lesser. Common and gregarious winter visitor all along Arabian Sea coast, frequenting sandy beaches, mangroves and rocky headlands.

1 **NORTHERN LAPWING** *Vanellus vanellus* 28–31 cm

a b ADULT NON-BREEDING Black crest, white (or buff) and black face pattern, black breast-band and dark green upperparts. Shows all-dark upperwing, and whitish rump and blackish tail-band in flight. Has very broad, rounded wing-tips. Distinctive slow-flapping flight with rather erratic wingbeats. Common winter visitor in northern areas, favouring grasslands on edges of lakes or seepage areas, and along the Indus south to Dadu District. Rare in southern Sind. Flocks of 200–300 winter around the shores of Ghauspur Lake.

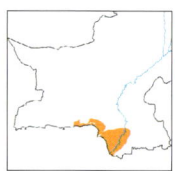

2 **YELLOW-WATTLED LAPWING**
Vanellus malabaricus 26–28 cm

a b ADULT Yellow wattles and legs. White supercilium, dark cap and brown breast-band. Wing and tail pattern much as Red-wattled. Call is a strident *chee-eet* and a hard *tit-tit-tit*. Birds begin territorial displays in early March, calling *tee-whit-tee-whit* followed by rapid *te-wit-wit-wit-te-wit-wit-wit*. Summer breeding visitor from India, coming at end of February and confined to extreme southern Sind and Las Bela. Prefers dry fallow fields and gravelly plains, avoiding marshy areas.

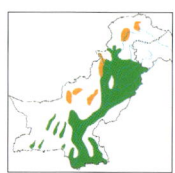

3 **RED-WATTLED LAPWING** *Vanellus indicus* 32–35 cm

a ADULT and **b** JUVENILE Black cap and breast, red bill with black tip, and yellow legs. Wing and tail pattern much as Yellow-wattled. Call is an agitated *did he do it, did he do it*. Abundant resident throughout the Indus basin, occurring wherever there is open ground adjacent to wetlands; largely absent from dry mountainous areas. Addicted to foraging in freshly irrigated cropland. Often breeds on banks of larger canals.

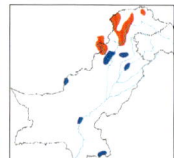

4 **SOCIABLE LAPWING** *Vanellus gregarius* 27–30 cm

a b ADULT BREEDING and **c** ADULT NON-BREEDING Dark cap, with white supercilia that join at nape. Adult breeding has yellow wash to sides of head, and black-and-maroon patch on belly. Adult non-breeding and immature have duller head pattern, white belly and streaked breast. Dry fallow fields, scrub desert and wetland margins. Considered 'fairly common' at the beginning of the 20th century, but probably now rare in Pakistan, with recent records of small groups from NWFP, northern Punjab and Thar Desert. Globally threatened (Critically Endangered).

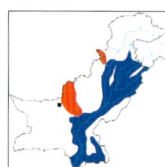

5 **WHITE-TAILED LAPWING** *Vanellus leucurus* 26–29 cm

a b ADULT and **c** JUVENILE Blackish bill, and very long yellow legs. Plain head. Tail all white, lacking black band of other *Vanellus* lapwings. Territorial display, a rapid *ti-toowhit-te-toowhit* call, rising to a crescendo before volplaning to earth again. Freshwater marshes and marshy lake edges. Common winter visitor throughout wetlands of Punjab and Sind, especially East Nara, with a few pairs breeding on shores of Zangi Nawar Lake in suitable water conditions.

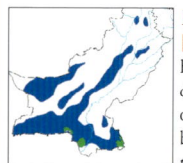

1 **CREAM-COLOURED COURSER** *Cursorius cursor*   21–24 cm
a **b** ADULT and **c** JUVENILE Pale sandy upperparts and underparts. Adult has sandy-rufous forehead, grey nape and pale lores. Juvenile has buffish crown and dark scaling on upperparts. Adapted to pure sand-dune desert, clay flats and stony open slopes. Prefers vicinity of herders' settlements, attracted by Scarabidae dung beetles. Uncommon resident in southern Sind and Mekran, with local migration throughout Baluchistan and west side of Indus, especially in winter, wherever there are open, treeless plains.

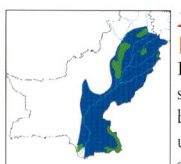

2 **INDIAN COURSER** *Cursorius coromandelicus*   23 cm
a **b** ADULT Grey-brown upperparts and orange underparts, with dark belly. Has chestnut crown, prominent white supercilium, and dark eye-stripe. In flight, shows white band across uppertail-coverts; underwing very dark, and wings broad with rounded wing-tips. Juvenile has brown barring on chestnut-brown underparts. Scarce resident, subject to nomadism, especially after monsoon. Often sympatric with Cream-coloured, and better adapted to fallow fields and desiccating margins of wetlands. Largely absent west of the Indus.

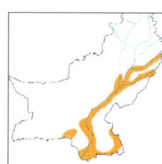

3 **COLLARED PRATINCOLE** *Glareola pratincola*   16–19 cm
a **b** **c** ADULT BREEDING, **d** ADULT NON-BREEDING and **e** JUVENILE Very similar to Oriental Pratincole (see Appendix), and best told from this species in all plumages by white trailing edge to secondaries (although this can be difficult to see). Adult has more pronounced fork to tail than in Oriental, with tail-tip reaching tips of closed wings at rest. Very graceful, feeding mainly by hawking insects on the wing in tern-like manner, at times high in the sky. Summer breeding visitor, wintering in East Africa. Common in lower Sind and rare in Punjab. Always in vicinity of desiccating margins of larger lakes, and salt-marsh flats along the coast, and in Badin on the borders of the Rann of Kutch.

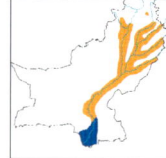

4 **SMALL PRATINCOLE** *Glareola lactea*   16–19 cm
a **b** **c** ADULT BREEDING and **d** ADULT NON-BREEDING Small size, with sandy-grey coloration and shallow fork to tail. In flight, shows white panel across secondaries, blackish underwing-coverts and black tail-band. Also hawks insects on the wing, often in large groups. Noisy during breeding season, calling *ke-ter-rick-ke-terrick* and *tirrittirrit*. Common summer visitor and breeder, along Indus river, larger lakes, and occasionally East Nara. Breeds colonially, preferring the desiccating margins of wetlands.

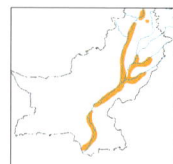

5 **INDIAN SKIMMER** *Rynchops albicollis*   40 cm
a ADULT and **b** JUVENILE Large, drooping orange-red bill. Juvenile has whitish fringes to upperparts. Flies close to the surface of the water, skimming the surface with bill open. Rare summer breeding visitor, confined to sandbars along the Indus and its main tributaries, arriving around the Indus delta from mid-February through to April. Globally threatened (Vulnerable).

JAEGERS AND LARGE GULLS, PLATE 32

1 **POMARINE JAEGER** *Stercorarius pomarinus* 56 cm
a ADULT DARK MORPH BREEDING, **b** ADULT PALE MORPH BREEDING, **c** ADULT PALE MORPH NON-BREEDING, **d** JUVENILE PALE MORPH and **e f** JUVENILE DARK MORPH Larger and stockier than Parasitic Jaeger, with heavier bill and broader-based wings. Adult breeding has long, broad central tail feathers twisted at end to form swollen tip (although tips can be broken off). Occurs in both pale and dark morphs. Adult non-breeding (pale morph) has indistinct cap, and barring to breast and upper- and undertail-coverts; uniform dark underwing-coverts distinguish it from birds in first- and second-winter plumage. Broader round-tipped central tail feathers best distinction from Parasitic. Juvenile variable, typically dark brown with broad pale barring on uppertail-, undertail- and underwing-coverts; head, neck and underparts never appear rufous-coloured as on some juvenile Parasitic (although other Parasitic appear virtually identical to Pomarine); note also the second pale crescent at base of primary coverts on underwing on most birds (in addition to pale base to underside of primaries), diffuse vermiculations (never streaking) on nape and neck, and blunt-tipped or almost non-existent projection of central tail feathers (more prominent and pointed in juvenile Parasitic). Scarce visitor attracted by feeding tern flocks along coastal waters, mostly in winter months. Majority of individuals are non-breeding immatures.

2 **PARASITIC JAEGER** *Stercorarius parasiticus* 45 cm
a ADULT DARK MORPH BREEDING, **b** ADULT PALE MORPH BREEDING, **c** ADULT PALE MORPH NON-BREEDING, **d** JUVENILE PALE MORPH, **e** JUVENILE INTERMEDIATE MORPH and **f** JUVENILE DARK MORPH Smaller and more lightly built than Pomarine Jaeger, with slimmer bill and narrower-based wings. Adult breeding has pointed tip to elongated central tail feathers. Occurs in both pale and dark morphs. Adult non-breeding is as Pomarine but has more pointed tail-tip. Juvenile more variable than juvenile Pomarine, ranging from grey and buff with heavy barring to completely blackish brown, and many have rusty-orange to cinnamon-brown cast to head and nape (not found on Pomarine); except for all-dark juveniles, further distinctions from Pomarine are dark streaking on head and neck, and pale tips to primaries. Frequent visitor to coastal waters, mainly of non-breeding immatures, from post-monsoon to spring, with a majority of dark morph birds.

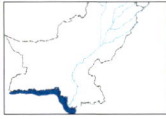

3 **HEUGLIN'S GULL** *Larus heuglini* 58–65 cm
a ADULT NON-BREEDING, **b** 1ST-WINTER and **c** 2ND-WINTER Darkest large gull of region. Adult has darker grey upperparts than Caspian and Pallas's gulls. (See account for Caspian Gull for further details on differentiation from that species.) Common winter visitor to coastal areas, only occasionally sighted inland on passage.

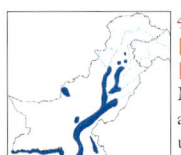

4 **CASPIAN GULL** *Larus cachinnans* 55–65 cm
a ADULT NON-BREEDING, **b** 1ST-WINTER, **c** 2ND-WINTER *L. c. cachinnans* and **d** ADULT NON-BREEDING *L. c. barabensis* Much larger and broader-winged than Mew Gull. More elongated appearance structurally, with smaller head, longer parallel-edged bill and longer legs, compared with Heuglin's. Adult has paler grey upperparts than Heuglin's, and head whiter and less heavily marked than Heuglin's in non-breeding plumage. First-winter told from Heuglin's by paler inner primaries, paler underwing-coverts with dark barring, and narrower tail-bar. Brown mottling on mantle best distinction from first-winter Pallas's. Second-year has paler grey mantle than second-year Heuglin's; diffusely barred tail, dark greater-covert bar, and lack of distinct mask help separate it from first-year Pallas's. Taxonomic position of *L. c. barabensis* is unclear and recent genetic analysis suggests it is more closely related to Heuglin's than Caspian. *L. c. barabensis* is a shade darker on the upperparts (in adult plumage) than *L. c. cachinnans*, and has a more rounded head, and a shorter bill and legs. Identification of 'large white-headed gulls' in the subcontinent is problematic and requires great caution. Less common than Heuglin's. Recorded inland and along Arabian Sea coast.

5 **PALLAS'S GULL** *Larus ichthyaetus* 69 cm
a ADULT BREEDING, **b** ADULT NON-BREEDING, **c** 1ST-WINTER and **d** 2ND-WINTER Angular head with gently sloping forehead, crown peaking behind eye. Bill large, 'dark-tipped', with bulging gonys. Adult breeding has black hood with bold white eye-crescents, and distinctive wing pattern. Adult non-breeding has largely white head with variable black mask. First-winter has grey mantle and scapulars; told from second-winter Caspian by head pattern (see account for Caspian Gull), absence of dark greater-covert bar, and more pronounced dark tail-band. Second-winter has largely grey upperwing, with dark lesser-covert bar and extensive black on primaries and primary coverts. Frequent winter visitor. Occasionally on larger lakes, more commonly on sea coast. Up to 30 winter on Khinjir Lake.

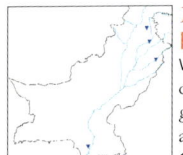

1 MEW GULL *Larus canus* — 43 cm
a ADULT NON-BREEDING, **b** 1ST-WINTER, **c** 1ST-SUMMER and **d** 2ND-WINTER Smaller and daintier than Caspian, with shorter and finer bill. Adult has darker grey mantle than Caspian, with more black on wing-tips; bill yellowish green, with dark subterminal band in non-breeding plumage, and dark iris. Head and hind-neck heavily marked in non-breeding (unlike adult non-breeding Caspian). First-winter/first-summer have uniform grey mantle; unbarred greyish greater coverts forming mid-wing panel, narrow black subterminal tail-band, and well-defined dark tip to greyish/pinkish bill are best distinctions (in addition to structural differences) from second-year Caspian (which also has grey mantle). Second-winter has black on primary coverts. Rare visitor to lakes and large rivers.

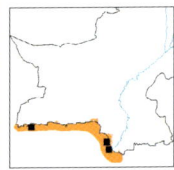

2 SOOTY GULL *Larus hemprichii* — 45 cm
a ADULT BREEDING, **b** 1ST-WINTER and **c** 2ND-WINTER Dark upperwing and underwing. Heavy, long, two-toned bill. Adult has brown hood, and greyish-brown mantle and upperwing; bill yellow with black-and-red tip. First-winter and second-winter have rather uniform brownish head and breast, and dark tail-band; bill initially greyish with black tip, becoming similar to adult during second winter. Common summer visitor March to October, with a few breeding in sand-dune areas along the coast, June onwards. Major breeding colony formerly on Astola Island, now extirpated due to introduced rats.

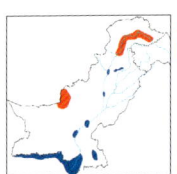

3 BROWN-HEADED GULL *Larus brunnicephalus* — 42 cm
a ADULT BREEDING, **b** ADULT NON-BREEDING and **c** 1ST-WINTER Larger than Black-headed, with more rounded wing-tips and broader bill. Adult has broad black wing-tips (broken by white 'mirrors') and white patch on outer wing; underside to primaries largely black; iris pale yellow (rather than brown as in adult Black-headed). In breeding plumage, hood paler brown than Black-headed's. First-winter has broad black wing-tips. Very common winter visitor, mainly along coastal waters, with smaller numbers on larger inland lakes and barrage headponds. Particularly common in Indus delta creeks, with roosting flocks of 100–200. Departs late April to early May.

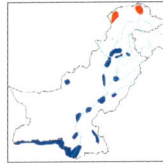

4 BLACK-HEADED GULL *Larus ridibundus* — 38 cm
a ADULT BREEDING, **b** ADULT NON-BREEDING and **c** 1ST-WINTER White 'flash' on primaries of upperwing. In non-breeding and first-winter, bill tipped black and head largely white with dark ear-covert patch. Abundant winter visitor to inland lakes, barrage headponds and sea coast. Strong spring passage through Baluchistan (e.g. over 3000 on Khushdil Khan Lake in March) and in northern areas. Departs by first week of April.

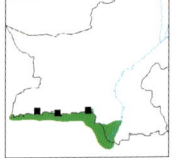

5 SLENDER-BILLED GULL *Larus genei* — 43 cm
a ADULT BREEDING, **b** ADULT NON-BREEDING and **c** 1ST-WINTER Head white throughout year, although may show grey ear-covert spot in winter. Gently sloping forehead and longish neck. Iris pale, except in juvenile. Adult has variable pink flush on underparts. Abundant winter visitor to sea coast and, post-monsoon, to coastal lagoons and lakes. Most of population breeds extralimitally in Kazakhstan but present throughout the year and, depending on water conditions, colonial breeding, occurs in inland wetlands along Mekran coast.

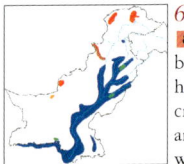

6 GULL-BILLED TERN *Gelochelidon nilotica* — 35–38 cm
a b ADULT BREEDING and **c** ADULT NON-BREEDING Stout, gull-like black bill and gull-like appearance. Grey rump and tail concolorous with back. Black half-mask in non-breeding and immature plumages. Juvenile has sandy tinge to crown and mantle, and dark subterminal markings to scapular and wing-coverts and tertials. Breeding season calls are a *ker-wick-wick ker-wick-ker-wick*. Common winter migrant, with a smaller resident population, breeding when water levels are suitable on sandbanks along the Indus and its tributaries, and on brackish lagoons in Las Bela. Prefers to hunt over mudbanks and canal and roadside seepage pools, picking its food off the surface.

LARGE AND MEDIUM-SIZED TERNS, PLATE 34

1 CASPIAN TERN *Sterna caspia* 47–54 cm
a ADULT BREEDING, **b** ADULT NON-BREEDING and **c** JUVENILE Large size with huge red bill. Black cap and crest are streaked with white in non-breeding and immature plumages. Juvenile has dark subterminal markings to coverts and tertials. Contact calls are a rasping *e-erragh-e-erragh*. Common resident along Arabian Sea coast, with breeding under suitable water conditions inland of Mekran coast, and on remote islands east of the Indus delta. Migrates along the Indus in winter and into larger inland lakes.

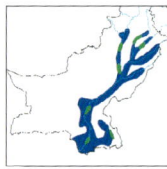

2 RIVER TERN *Sterna aurantia* 38–46 cm
a b ADULT BREEDING, **c** ADULT NON-BREEDING and **d** JUVENILE Adult breeding has orange-yellow bill, black cap, greyish-white underparts, and long tail. Large size, stocky appearance and stout yellow bill (with dark tip) help separate adult non-breeding and immature from Black-bellied. Exclusively fluvial common resident, breeding on sandbars along major rivers, but wandering widely to forage over inland lakes and larger irrigation canals. Virtually absent from hilly tracts.

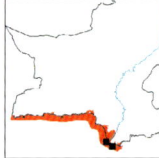

3 LESSER CRESTED TERN *Sterna bengalensis* 35–37 cm
a b ADULT BREEDING, **c** ADULT NON-BREEDING, **d** 1ST-WINTER and **e** JUVENILE Orange to orange-yellow bill, smaller and slimmer than Great Crested's, with paler grey upperparts (adult) and usually less boldly patterned upperwing (immatures). Common all along Arabian Sea coast. A marine species, breeding in small numbers on remote islands in the Indus delta. Much less numerous than both Sandwich and Great Crested terns inshore. Flocks observed fishing 24 km offshore.

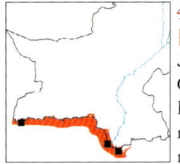

4 GREAT CRESTED TERN *Sterna bergii* 46–49 cm
a b ADULT BREEDING, **c** ADULT NON-BREEDING, **d** 1ST-WINTER and **e** JUVENILE Lime-green to cold-yellow bill. Larger and stockier than Lesser Crested, with darker grey upperparts (adult) or darker and usually more strongly patterned upperwing (immatures). Common all along Arabian Sea coast: a marine species. Prefers to forage over open seas rather than Indus creeks. Small numbers breed on remote islands in the Indus delta. The former major breeding colony on Astola Island now extirpated due to the introduction of rats.

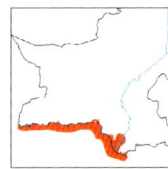

5 SANDWICH TERN *Sterna sandvicensis* 36–41 cm
a b ADULT BREEDING, **c** ADULT NON-BREEDING and **d** 1ST-WINTER Slim black bill with yellow tip, and more rakish appearance than Gull-billed. White rump and tail contrast with greyer back. U-shaped black crest in non-breeding and first-winter/first-summer plumages. Juvenile more heavily marked than juvenile Gull-billed, and has black rear crown and nape. Coasts, tidal creeks and open sea. In recent years, the Sandwich has been the commonest tern along the Arabian Sea coast. Recorded from October to June, but breeding birds migrate to the Caspian Sea region.

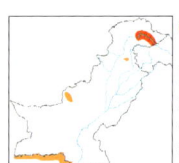

6 COMMON TERN *Sterna hirundo* 31–35 cm
a b ADULT BREEDING, **c** ADULT NON-BREEDING and **d** JUVENILE In breeding plumage has orange-red bill with black tip, pale grey wash to underparts, and long tail streamers. Has dark bill, whitish forehead and dark lesser-covert bar in non-breeding plumage. Birds in full breeding plumage are often seen together with others in worn plumage and all-black bills in April/May. Uncommon visitor to lakes, large rivers and sea coast mainly summer months. Also scattered records in all winter months. Birds breeding in Tibet and Ladakh are regularly seen in early May following the Indus and Shyok rivers in Baltistan.

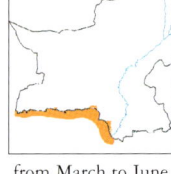

7 WHITE-CHEEKED TERN *Sterna repressa* 32–34 cm
a b ADULT BREEDING, **c** ADULT NON-BREEDING and **d** JUVENILE Smaller than Common, with darker grey upperparts, and uniform grey rump and tail concolorous with back; underwing has darker trailing edge and pale central panel. In breeding plumage, darker grey on underparts than Common, and with white cheeks; from adult breeding Whiskered by longer bill, paler grey underparts and longer tail streamers. Common summer visitor all along Arabian Sea coast, foraging in flocks from March to June. Smaller numbers recorded offshore October to December. Breeds extralimitally.

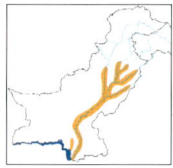

1 **LITTLE TERN** *Sterna albifrons* 22–24 cm
a b ADULT BREEDING, **c** ADULT NON-BREEDING and **d** JUVENILE Fast flight with rapid wingbeats, and narrow-based wings. Frequently hovers before plunge-diving. Adult breeding has white forehead and black-tipped yellow bill. Adult non-breeding has blackish bill, black mask and nape-band, and dark lesser-covert bar. Common winter visitor along sea coast and uncommon summer breeder on major rivers of Indus basin.

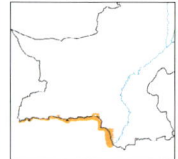

2 **SAUNDERS'S TERN** *Sterna saundersi* 23 cm
a b ADULT BREEDING and **c** 1ST-WINTER Adult breeding as Little Tern, but more rounded white forehead patch (lacking short white supercilium), shorter reddish-brown to brown legs (orange on Little), broader black outer edge to primaries, and grey rump and centre of tail. There are no sure features for separating other plumages from Little, although darker grey upperparts including rump and dark bar on secondaries may be useful. Common summer breeding visitor along Arabian Sea coast. A maritime species, nesting in small dispersed colonies. Suffers much predation from feral dogs and House Crows.

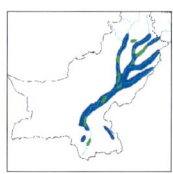

3 **BLACK-BELLIED TERN** *Sterna acuticauda* 33 cm
a b ADULT BREEDING, **c** ADULT NON-BREEDING and **d** JUVENILE Much smaller than River Tern, with orange bill (with variable black tip) in all plumages. Adult breeding has black belly and vent. Adult non-breeding and juvenile have white underparts, and black mask and streaking on crown. Marshes, lakes and rivers. Uncommon resident, formerly more numerous. Suffers much predation while breeding on sandbars of major rivers.

4 **BRIDLED TERN** *Sterna anaethetus* 30–32 cm
a b ADULT BREEDING, **c** ADULT NON-BREEDING and **d e** JUVENILE In breeding plumage, white forehead patch extends over eye as broad white supercilium, and has brownish-grey mantle and wings. Juvenile has greyish-white crown, dark mask and white forehead and supercilium (shadow-pattern of adult), buffish fringes to mantle and wing-coverts, and brownish patch on side of breast. Exclusively pelagic in its foraging. Breeds in Persian Gulf region. Common only along western coast of Mekran.

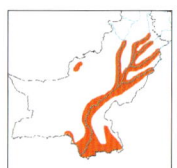

5 **WHISKERED TERN** *Chlidonias hybridus* 23–25 cm
a b ADULT BREEDING, **c d** ADULT NON-BREEDING and **e** JUVENILE In breeding plumage, white cheeks contrast with grey underparts; lacks greatly elongated outer tail feathers. In non-breeding and juvenile, distinguished from White-winged by larger bill, grey rump concolorous with back and tail, and different patterning of black on head. Very common spring and autumn passage migrant, inhabiting inland lakes, rivers and flooded paddy-fields. Breeds extralimitally in Kashmir and northern India.

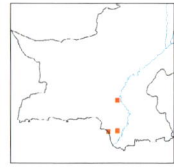

6 **WHITE-WINGED TERN** *Chlidonias leucopterus* 20–23 cm
a ADULT BREEDING, **b c** ADULT NON-BREEDING and **d** JUVENILE In breeding plumage, black head and body contrast with pale upperwing-coverts, and has black underwing-coverts. In non-breeding and juvenile, smaller bill, whitish rump contrasting with grey tail, and different patterning of black on head are distinctions from Whiskered. Rare but regular visitor only on spring passage, occurring exclusively over freshwater marshes, lakes and rivers.

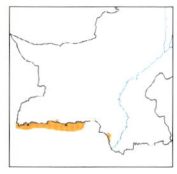

7 **BROWN NODDY** *Anous stolidus* 42 cm
a b ADULT and **c d** JUVENILE Brown, with coverts paler than flight feathers, and pale panel across upperwing-coverts. Bill stouter, proportionately shorter and noticeably down-curved compared with those of Black Noddy and Lesser Noddy (although these species have not been recorded in Pakistan). Juvenile has browner forehead and crown. A purely pelagic tropical ocean species, occurring as rare visitor off Baluchistan coast.

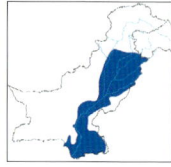

1 **OSPREY** *Pandion haliaetus*   55–58 cm
a b ADULT Long wings, typically angled at carpals, and short tail. Has whitish head with black stripe through eye, white underbody and underwing-coverts, and black carpal patches. Frequently hovers over water when fishing. Common and widespread winter visitor on all larger inland lakes and barrage headponds. More plentiful on seashores and coastal waters. A few non-breeders over-summer.

2 **BLACK-SHOULDERED KITE** *Elanus caeruleus*   31–35 cm
a b c ADULT and **d** JUVENILE Small size. Grey and white with black 'shoulders'. Flight buoyant, with much hovering. Juvenile has brownish-grey upperparts with pale fringes, with less distinct shoulder-patch. Breeding male calls repeatedly, an upward-inflected whistle, *toowhit-toowhit*. Commonest raptor, resident throughout the Indus basin, extending into pure desert, but avoiding mountainous areas. Prefers lightly wooded country, avoiding forest plantations.

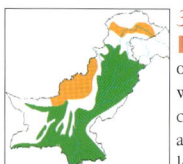

3 **BLACK KITE** *Milvus migrans*   55–68.5 cm
a b c ADULT and **d e** JUVENILE Shallow tail-fork. Much manoeuvring of arched wings and twisting of tail in flight. Dark rufous brown, with variable whitish crescent at primary bases on underwing, and pale band across median coverts on upperwing. Juvenile has broad whitish or buffish streaking on head and underparts. Common resident population (ssp. *govinda*) throughout lowlands. Scarce migrant summer breeding (ssp. *migrans*) in Baluchistan and NWFP. Winter visitor (ssp. *lineatus*), favouring major wetlands.

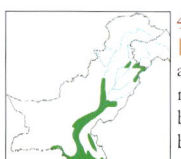

4 **BRAHMINY KITE** *Haliastur indus*   48 cm
a b ADULT and **c d** JUVENILE Small size and kite-like flight. Wings usually angled at carpals. Tail rounded. Adult mainly chestnut, with white head, neck and breast. Juvenile mainly brown, with pale streaking on head, mantle and breast, large pale patches at bases of primaries on underwing, and cinnamon-brown tail. Breeding males call a quavering mammalian-like *mee-ah*. Thinly distributed resident throughout Indus river and major tributaries. Very common year-round in southern Sind and mangrove coastal areas.

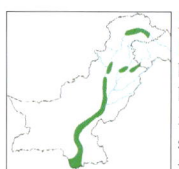

5 **PALLAS'S FISH EAGLE** *Haliaeetus leucoryphus*   76–84 cm
a b ADULT and **c d** JUVENILE Soars and glides with wings flat. Long, broad wings and protruding head and neck. Adult has pale head and neck, dark brown upperwing and underwing, and mainly white tail with broad black terminal band. Juvenile less bulky, looks slimmer-winged, longer-tailed and smaller-billed than juvenile White-tailed; has dark mask, pale band across underwing-coverts, pale patch on underside of inner primaries, all-dark tail, and pale crescent on uppertail-coverts. During breeding both sexes duet, a far-carrying *gwark-gwark*, rising and falling in pitch. Large rivers and lakes. Severely declining resident. Population estimated (1972) at less than 40 pairs. Traditionally sedentary, using same tree nest annually; many abandoned due to human disturbance. Globally threatened (Vulnerable).

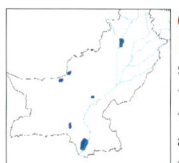

6 **WHITE-TAILED EAGLE** *Haliaeetus albicilla*   70–90 cm
a b ADULT and **c d** JUVENILE Huge, with broad parallel-edged wings, short wedge-shaped tail and protruding head and neck. Soars and glides with wings level. Adult has large yellow bill, pale head and white tail. Juvenile has whitish centres to tail feathers, pale patch on axillaries, and variable pale band across underwing-coverts. Catches fish and waterfowl by flying low over the water surface. A rare winter visitor, migrating down the Indus (predominantly sub-adult birds). Attracted to wetlands with plenty of waterbirds.

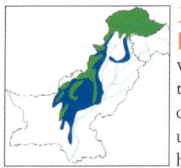

1 **LAMMERGEIER** *Gypaetus barbatus* 100–115 cm
a b ADULT and **c d** IMMATURE Huge size, long and narrow pointed wings, and large wedge-shaped tail. Adult has blackish upperparts, wings and tail, and cream or rufous-orange underparts contrasting with black underwing-coverts. Immature has blackish head and neck and grey-brown underparts. Has unique habit of splitting bones for the marrow by dropping them from a great height. Widespread but scarce resident in mountain regions, from arid hills of central Baluchistan, through green alpine pastures and lightly forested areas up to the far north of the Karakorams. Now only a casual visitor to Murree Hills and, in winter, to Sind Kohistan.

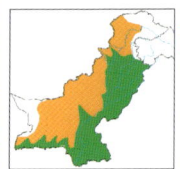

2 **EGYPTIAN VULTURE** *Neophron percnopterus* 60–70 cm
a b c ADULT, **d e** IMMATURE and **f** JUVENILE Small vulture with long, pointed wings, small and pointed head, and wedge-shaped tail. Adult mainly dirty white, with bare yellowish face and black flight feathers. Juvenile blackish brown with bare grey face. With maturity, tail, body and wing-coverts become whiter and face yellower. Now uncommon resident, but widespread throughout Indus basin and breeding in summer in Baluchistan and northern Himalayas (map shows former distribution). Well adapted to desert areas, and preferring outskirts of towns, especially refuse dumps. Globally threatened (Endangered).

3 **WHITE-RUMPED VULTURE** *Gyps bengalensis* 75–85 cm
a b c ADULT and **d e f** JUVENILE Smallest of the *Gyps* vultures. Adult mainly blackish, with white rump and back, and white underwing-coverts. Key features of juvenile are dark brown coloration, streaking on underparts and upperwing-coverts, dark rump and back, whitish head and neck, and all-dark bill. In flight, underbody and underwing-coverts distinctly darker than on Eurasian Griffon and Indian Vulture. Juvenile is similar in coloration to juvenile Himalayan Griffon, but much smaller and less heavily built, with narrower-looking wings and shorter tail, underparts less heavily streaked, and lacking prominent streaking on mantle and scapulars. Now uncommon resident in the plains, breeding colonially in tree plantations or riverine forest (map shows former distribution). Population decimated by widespread use of anti-inflammatory drug diclofenac on livestock, which, if scavenged, causes renal failure. Now close to extirpation in Pakistan. Globally threatened (Critically Endangered).

4 **INDIAN VULTURE** *Gyps indicus* 80–95 cm
a b ADULT and **c** JUVENILE Adult has sandy-brown body and upperwing-coverts, blackish head and neck with sparse white down on hind-neck, white downy ruff, and yellowish bill and cere, and lacks pale streaking on underparts; in flight, lacks broad whitish band across median underwing-coverts shown by Eurasian Griffon, and has whiter rump and back. Juvenile has feathery buff neck-ruff, dark bill and cere with a pale culmen, and whitish down on the head and neck; distinguished from juvenile Eurasian Griffon by pale culmen, darker brown upperparts with more pronounced pale streaking, and paler and less rufescent coloration to streaked underparts; best distinctions from juvenile White-rumped are paler and less clearly streaked underparts, paler upper- and underwing-coverts, and whitish rump and back. A tiny population is resident and breeding on cliffs in extreme southwest Tharparkar District. Because Kohli and Bheel Hindu peasants in that area lack veterinary help, and are not thought to be using the drug diclofenac, which kills vultures, this population may still survive. Globally threatened (Critically Endangered).

VULTURES, PLATE 38

1 HIMALAYAN GRIFFON *Gyps himalayensis* 115–125 cm

a b c ADULT and **d e f** JUVENILE Larger than Eurasian Griffon, with broader body and slightly longer tail. Wing-coverts and body pale buffish-white, contrasting strongly with dark flight feathers and tail, and ruff is buffish; underparts lack pronounced streaking; legs and feet pinkish with dark claws, and has yellowish bill and cere. Juvenile has brown feathered ruff, with bill and cere initially black, and has dark brown body and upperwing-coverts boldly and prominently streaked with buff (wing-coverts almost concolorous with flight feathers), and back and rump also dark brown; streaked upperparts and underparts and pronounced white banding across underwing-coverts are best distinctions from Cinereous Vulture; very similar in plumage to juvenile White-rumped Vulture, but much larger and more heavily built, with broader wings and longer tail, underparts more heavily streaked, and streaking on mantle and scapulars. Formerly common throughout northern mountainous regions and more sedentary than Eurasian Griffon, descending as far as the Murree Hills in winter. Absent from Baluchistan. Present status unclear, due to concerns about impact of diclofenac (see account of White-rumped Vulture).

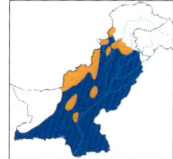

2 EURASIAN GRIFFON *Gyps fulvus* 95–105 cm

a b c ADULT and **d e f** JUVENILE Larger and stockier than Indian, with stouter bill. Key features of adult are yellowish bill with blackish cere, whitish head and neck, fluffy white ruff, rufescent-buff upperparts, rufous-brown underparts and thighs with prominent pale streaking, and dark grey legs and feet; rufous-brown underwing-coverts usually show prominent whitish banding (especially across medians) (see accounts for Indian Vulture and Himalayan Griffon for comparison). Immature has richer rufous-brown on upperparts and upperwing-coverts (with prominent pale streaking) than adult; has rufous-brown feathered neck-ruff, more whitish down covering grey head and neck, blackish bill and dark iris (pale yellowish brown in adult). Formerly common and widespread, breeding colonially in drier mountain ranges of Baluchistan, the Punjab Salt Range and foothills of NWFP. Winter migrant throughout the plains and eastern desert borders. Present status unknown, due to concerns about impact of diclofenac (see account of White-rumped Vulture).

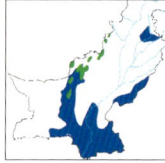

3 CINEREOUS VULTURE *Aegypius monachus* 100–110 cm

a b ADULT and **c** JUVENILE Very large vulture with broad, parallel-edged wings. Soars with wings flat (wings usually held in shallow V in *Gyps* species). At a distance appears typically uniformly dark, except for pale areas on head and bill. Adult blackish brown with paler brown ruff; may show paler band across greater underwing-coverts, but underwing darker and more uniform than on *Gyps* species. Juvenile blacker and more uniform than adult. Scarce resident, breeding in high-altitude juniper forest, migrating in winter throughout plains and desert areas. Roosts gregariously on cliff escarpments in winter. Badly affected by trapping for the zoo trade.

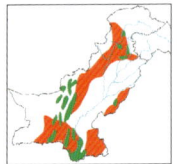

1 **SHORT-TOED SNAKE EAGLE** *Circaetus gallicus* 62–67 cm
a b c ADULT PALE PHASE and **d** ADULT DARK PHASE Long and broad wings, pinched in at base, and rather long tail. Head broad and rounded. Soars with wings flat or slightly raised; frequently hovers. Pattern variable, often with dark head and breast, barred underbody, dark trailing edge to underwing, and broad subterminal tail-band; can be very pale on underbody and underwing. Uncommon resident, sparsely distributed throughout Pakistan, except irrigated Indus plains and northern Himalayas. More frequent in southern Sind (as many as 6 dispersed individuals can be in view in Thatta District).

2 **BLACK EAGLE** *Ictinaetus malayensis* 69–81 cm
a b ADULT and **c d** JUVENILE In flight, has distinctive wing shape and long tail. Flies with wings raised in V, with primaries upturned. At rest, long wings extend to tip of tail. Adult dark brownish black, with striking yellow cere and feet; in flight, shows whitish barring on uppertail-coverts, and faint greyish barring on tail and underside of remiges. Juvenile has dark-streaked buffish head, underparts and underwing-coverts. Hunts by sailing buoyantly and slowly over the canopy, sometimes weaving in and out of treetops. Rare and irregular visitor to forests of the Murree Hills in summer months. No breeding records since end of 19th century.

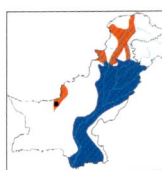

3 **EURASIAN MARSH HARRIER** *Circus aeruginosus* 48–58 cm
a b c MALE, **d e f** FEMALE and **g** JUVENILE Broad-winged and stocky. As with other harriers, glides and soars with wings in noticeable V, quartering the ground a few metres above it, occasionally dropping to catch prey. Male has brown mantle and upperwing-coverts contrasting with grey secondaries/inner primaries; female mainly dark brown, except for cream on head and on leading edge of wing. Juvenile may be entirely dark. Marshes, lakes and grasslands. Common winter visitor, when females are predominant, with occasional breeding at Zangi Nawar Lake when water conditions are suitable. Most numerous in rice-growing areas. Twenty counted at Sadori Lake, Sanghar District, one November.

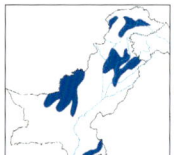

1 **HEN HARRIER** *Circus cyaneus* 44–52 cm

a b c MALE and **d e f** FEMALE Comparatively broad-winged and stocky. Male has dark grey upperparts, extensive black wing-tips and lacks black secondary bars. Female has broad white band across uppertail-coverts and rather plain head pattern (usually lacking dark ear-covert patch). Juvenile has streaked underparts as female, but with rufous-brown coloration. Open country plains and hills. Scarce winter visitor, preferring main valleys in northern areas and outer foothill regions, with marked spring and autumn passage through central Baluchistan.

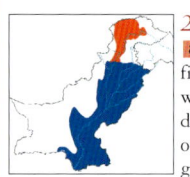

2 **PALLID HARRIER** *Circus macrourus* 40–48 cm

a b c MALE, **d e f** FEMALE and **g h** JUVENILE Slim-winged and fine-bodied, with buoyant flight. As with other harriers, glides and soars with wings in noticeable V, quartering the ground a few metres above it, occasionally dropping to catch prey. Folded wings fall short of tail-tip, and legs longer than on Montagu's. Male has pale grey upperparts, dark wedge on primaries, very pale grey head and underbody, and lacks black secondary bars. Female has distinctive underwing pattern: pale primaries, irregularly barred and lacking dark trailing edge, contrast with darker secondaries, which have pale bands that are narrower than on female Montagu's and taper towards body (although first-summer Montagu's more similar in this respect), and lacks prominent barring on axillaries. Typically, female has stronger head pattern than Montagu's, with more pronounced pale collar, dark ear-coverts and dark eye-stripe, and upperside of flight feathers darker and lacking banding; told from female Hen by narrower wings with more pointed hand, stronger head pattern, and patterning of underside of primaries. Juvenile has evenly barred primaries (lacking pronounced dark fingers), without dark trailing edge, and usually with pale crescent at base; head pattern more pronounced than Montagu's, with narrower white supercilium, more extensive dark ear-covert patch, and broader pale collar contrasting strongly with dark neck sides. Winter visitor and commonest harrier after Eurasian Marsh. Avoids mountainous regions, except on passage, and favours non-irrigated or desert tracts for hunting.

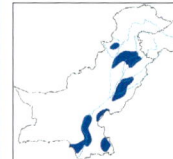

3 **MONTAGU'S HARRIER** *Circus pygargus* 43–47 cm

a b c MALE, **d e f** FEMALE and **g h** JUVENILE Folded wings reach tail-tip, and legs shorter than on Pallid. Male has black band across secondaries, extensive black on underside of primaries, and rufous streaking on belly and underwing-coverts. Female differs from female Pallid in distinctly and evenly barred underside to primaries, these with dark trailing edge, broader and more pronounced pale bands across secondaries, barring on axillaries, less pronounced head pattern, and distinct dark banding on upperside of remiges. Juvenile has unstreaked rufous underparts and underwing-coverts, and darker secondaries than female; differs from juvenile Pallid in having broad dark fingers and dark trailing edge to hand on underwing, and paler face with smaller dark ear-covert patch and less distinct collar. Scarce winter visitor, least common of harriers, occurring throughout Indus basin and preferring treeless open tracts for hunting.

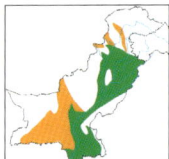

1 **SHIKRA** *Accipiter badius* 30–36 cm
a MALE, **b c** FEMALE and **d e** JUVENILE Adults paler than Besra (see Appendix) and Eurasian Sparrowhawk. Underwing pale, with fine barring on remiges, and slightly darker wing-tips. Male has pale blue-grey upperparts, indistinct grey gular stripe, fine brownish-orange barring on underparts, unbarred white thighs, and unbarred or only lightly barred central tail feathers. Upperparts of female are more brownish grey. Juvenile has pale brown upperparts, more prominent gular stripe, and streaked underparts; distinguished from juvenile Besra by paler upperparts and narrower tail barring, and from Eurasian Sparrowhawk by streaked underparts. Noisy in breeding season, repeating high-pitched *ti-whick-ti-whick*. Common resident throughout Sind and Punjab, preferring well-wooded areas. Summer breeder in northern Baluchistan, and in subtropical pine forest (*Pinus roxburghii*) in Himalayan foothills.

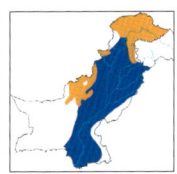

2 **EURASIAN SPARROWHAWK** *Accipiter nisus* 31–36 cm
a b MALE, **c d** FEMALE and **e** JUVENILE Upperparts of adult darker than those of Shikra, with prominent tail-barring, and strongly barred underwing. Uniform barring on underparts and absence of prominent gular stripe should separate it from Besra (see Appendix). Male has dark slate-grey upperparts and reddish-orange barring on underparts. Female is dark brown on upperparts, with dark brown barring on underparts. Juvenile has dark brown upperparts and barred underparts. Noisy in breeding season. Territorial calls are a drawn-out *kyew-kyew-kyew*, and warning calls are a chattering *ke-kek-kek*. Uncommon resident, and sparsely distributed winter visitor throughout Indus basin. Summer breeder in juniper forest, Baluchistan, and throughout Himalayas from Murree Hills north to Hunza and Baltistan.

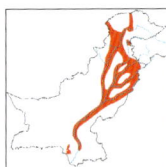

3 **NORTHERN GOSHAWK** *Accipiter gentilis* 50–61 cm
a b FEMALE and **c** JUVENILE Very large, with heavy, deep-chested appearance. Wings comparatively long, with bulging secondaries. Male has grey upperparts (greyer than in female Eurasian Sparrowhawk), white supercilium and finely barred underparts. Female is considerably larger, with browner upperparts. Juvenile has heavy streaking on buff-coloured underparts. Formerly regular migrant to well-wooded areas, arriving through Chitral and Gilgit main valleys. Now very rare due to constant trapping. Much favoured by Punjabi landowners for 'hawking'.

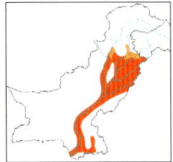

1 ORIENTAL HONEY-BUZZARD
Pernis ptilorhyncus   57–60 cm

a b c MALE and **d** FEMALE Long, broad wings and tail, narrow neck and small head with small crest. Soars with wings flat. Very variable in plumage; often shows dark moustachial stripe and gular stripe, and gorget of streaking across lower throat. Lacks dark carpal patch. Male has grey face, greyish-brown upperparts, 2 black tail-bands, usually 3 black bands across underside of remiges, and dark brown iris. Female has browner face and upperparts, 3 black tail-bands, 4 narrower black bands across remiges, and yellow iris. Uncommon and locally migratory, largely influenced by movement of Rock Bees and smaller Wild Honeybees. Summer breeder in northern Punjab, NWFP and subtropical pine zone. Mainly a winter visitor to lower Sind, everywhere favouring irrigated forest plantations and relict patches of riverine forest.

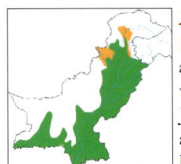

2 WHITE-EYED BUZZARD *Butastur teesa*   43 cm

a b c ADULT and **d e** JUVENILE Longish, rather slim wings, long tail, and buzzard-like head. Pale median-covert panel. Adult has black gular stripe, white nape-patch, barred underparts, dark wing-tips, and rufous tail; iris yellow. Juvenile has buffish head and breast streaked with dark brown, with moustachial and throat stripes indistinct or absent; rufous uppertail more strongly barred; iris brown. Noisy during breeding season, loudly repeated disyllabic scream, *ki-weeahr*. Dry open country, scrub and dry open forest. Very common resident throughout the Indus basin and foothill areas, extending into main desert tracts and across southern Baluchistan. Absent from extensive mountainous tracts and limited summer migration into subtropical pine zone.

3 COMMON BUZZARD *Buteo buteo*   51–56 cm

a b c d ADULT *B. b. japonicus* and **e** ADULT *B. b. refectus* Stocky, with broad wings and moderate-length tail. Soars with wings held in V shape. Variable; some very similar to Long-legged. *B. b. japonicus* typically has rather pale head and underparts, with variable dark streaking on breast and brown patch on belly/thighs; tail dark-barred grey-brown. *B. b. refectus* is dark brown to rufous brown, with variable amounts of white on underparts; tail dull brown with some dark barring, or uniform sandy brown. *B. b. vulpinus* (not illustrated) is extremely variable, and usually has rufous tail (and is similar to Long-legged; see that species). Frequent winter visitor all over Pakistan, less common in irrigated cultivated areas, preferring more open arid country, especially the subspecies *B. b. vulpinus*.

4 LONG-LEGGED BUZZARD *Buteo rufinus*   61 cm

a b c d ADULT Larger and longer-necked than Common, with longer wings and tail (appears more eagle-like); soars with wings in deeper V. Variable in plumage. Most differ from Common in having combination of paler head and upper breast, rufous-brown lower breast and belly, more uniform rufous underwing-coverts, more extensive black carpal patches, larger pale primary patch on upperwing, and unbarred pale orange uppertail. Rufous and black morphs are similar to some plumages of Common. Widespread, common winter visitor all over Pakistan, except in high mountainous tracts. A rare breeder in sparsely forested inner Himalayan regions. Less common in irrigated cultivated tracts, preferring arid scrub desert and stony foothills.

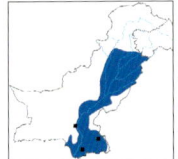

1 GREATER SPOTTED EAGLE *Aquila clanga* 65–72 cm

a b c ADULT, **d e f** JUVENILE and **g h** JUVENILE *'fulvescens'* Medium-sized eagle with rather short and broad wings, stocky head and short tail. Wings distinctly angled down at carpals when gliding, almost flat when soaring. Compared with Steppe Eagle, has less protruding head in flight, with shorter wings and less deeply fingered wing-tips; at rest, trousers are less baggy, and bill is smaller with rounded (rather than elongated) nostril and shorter gape; lacks adult Steppe's barring on underside of flight and tail feathers, and dark trailing edge to wing, and has a dark chin. Pale variant *'fulvescens'* distinguished from juvenile Imperial Eagle by structural differences, lack of prominent pale wedge on inner primaries on underwing, and unstreaked underparts. Juvenile has bold whitish tips to dark brown coverts. (See Appendix for comparison with Indian Spotted Eagle.) Vocal throughout year, the calls a high-pitched yelping *kyew-kyew*. Common winter visitor, preferring vicinity of large lakes, barrage headponds and large canals. Rare breeder in better wooded parts of Sind. Globally threatened (Vulnerable).

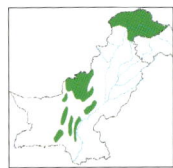

2 GOLDEN EAGLE *Aquila chrysaetos* 75–88 cm

a b c ADULT and **d e** JUVENILE Large eagle, with long and broad wings (with pronounced curve to trailing edge), long tail, and distinctly protruding head and neck. Wings clearly pressed forward and raised (with upturned fingers) in pronounced V when soaring. Adult has pale panel across upperwing-coverts, gold crown and nape, and two-toned tail. Juvenile has white base to tail and white patch at base of flight feathers. Scarce resident in high mountainous areas, from northern Baluchistan up to northern borders of Pakistan. A few sub-adults migrate to foothills in winter.

3 TAWNY EAGLE *Aquila rapax* 63–71 cm

a b c d e ADULT, **f g** JUVENILE and **h** SUB-ADULT Compared with Steppe, hand of wing does not appear so long and broad, tail is slightly shorter, and looks smaller and weaker at rest; gape-line ends level with centre of eye (extends to rear of eye in Steppe), and adult has yellowish iris. Differs from Greater Spotted in having more protruding head and neck in flight, baggy trousers, yellow iris, and oval nostril. Adult extremely variable, from dark brown through rufous to pale cream, and unstreaked or streaked with rufous or dark brown. Dark morph very similar to adult Steppe (which shows much less variation); distinctions include less pronounced barring and dark trailing edge on underwing, dark nape and dark chin. Rufous to pale cream Tawny is uniformly pale from uppertail-coverts to back, with undertail-coverts same colour as belly (contrast often apparent on similar species). Pale adults also lack prominent whitish trailing edge to wing, tip to tail and greater-covert bar (present on immatures of similar species). Characteristic, if present, is distinct pale inner-primary wedge on underwing. Juvenile also variable, with narrow white tips to unbarred secondaries; otherwise, as similar-plumaged adult. Many (possibly all) non-dark Tawny have distinctive immature/sub-adult plumage: dark throat and breast contrast with pale belly, and can show dark banding across underwing-coverts; whole head and breast may be dark. Uncommon resident, preferring desert tracts, but adapted to irrigated cultivation. Absent in high mountainous tracts, and scarce in Baluchistan.

4 STEPPE EAGLE *Aquila nipalensis* 76–80 cm

a b c ADULT, **d e f** JUVENILE and **g h** IMMATURE Broader and longer wings than those of Greater Spotted, with more pronounced and spread fingers, and more protruding head and neck; wings flatter when soaring, and less distinctly angled down at carpals when gliding. When perched, clearly bigger and heavier, with heavier bill and baggy trousers. Adult separated from adult Greater Spotted by underwing pattern (dark trailing edge, distinct barring on remiges, indistinct/non-existent pale crescents in carpal region), pale rufous nape-patch and pale chin. Juvenile has broad white bar across underwing, double white bar on upperwing, and white crescent across uppertail-coverts; prominence of bars on upperwing and underwing much reduced on older immatures. (See Appendix for comparison with Indian Spotted Eagle.) Winter migrant, when more plentiful than Tawny. Favours vicinity of large lakes and wetlands. A communal roost of 30–40 can be seen on escarpments south of Islamabad.

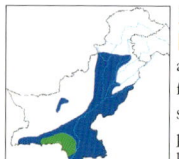

1 **IMPERIAL EAGLE** *Aquila heliaca* 72–83 cm

a b c ADULT and **d e f** JUVENILE Large, stout-bodied eagle with long and broad wings, longish tail, and distinctly protruding head and neck. Wings flat when soaring and gliding. Adult has almost uniform upperwing, small white scapular patches, golden-buff crown and nape, and two-toned tail. Juvenile has pronounced curve to trailing edge of wing, pale wedge on inner primaries, streaked buffish body and wing-coverts, uniform pale rump and back (lacking distinct pale crescent shown by other species except Tawny), and white tips to median and greater upperwing-coverts. A scarce winter visitor, preferring open plains and deserts, and avoiding high mountainous regions. Attracted to congregations of waterfowl. Occasional breeder in foothill country in early 20th century. No recent records. Globally threatened (Vulnerable).

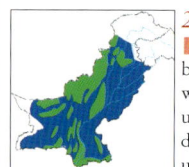

2 **BONELLI'S EAGLE** *Hieraaetus fasciatus* 65–72 cm

a b c ADULT and **d e f** JUVENILE Medium-sized eagle with long and broad wings, distinctly protruding head, and long, square-ended tail. Soars with wings flat. Adult has pale underbody and forewing, blackish band along underwing-coverts, whitish patch on mantle, and pale greyish tail with broad dark terminal band. Juvenile has ginger-buff to reddish-brown underbody and underwing-coverts (with variable dark band along greater underwing-coverts), uniform upperwing, and pale crescent on uppertail-coverts and patch on back. Widespread but scarce resident, preferring warmer, drier hill ranges, but breeding throughout Baluchistan and NWFP, except far northern cold mountainous regions.

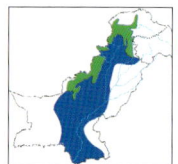

3 **BOOTED EAGLE** *Hieraaetus pennatus* 45–53 cm

a b c PALE MORPH and **d e** DARK MORPH Smallish eagle with long wings and long, square-ended tail. Glides and soars with wings flat or slightly angled down at carpal. Always shows white shoulder patches, pale median-covert panel, pale wedge on inner primaries, white crescent on uppertail-coverts, and greyish undertail with darker centre and tip. Head, body and wing-coverts whitish, brown or rufous, respectively, in pale, dark and rufous morphs. Uncommon resident, breeding (1200–3000 m) in inner mountain ranges of northwestern Baluchistan, NWFP, and from Chitral to Gilgit. Locally migratory to adjacent foothills in winter, rare in the plains. For nesting, prefers drier mountains with juniper, Deodar or subtropical pine forest.

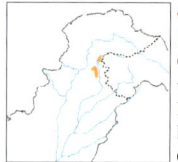

4 **MOUNTAIN HAWK EAGLE** *Spizaetus nipalensis* 70–72 cm

a b ADULT and **c d** JUVENILE Prominent crest. Wings broader than on Changeable Hawk Eagle (N.B. this species has not been recorded in Pakistan), with more pronounced curve to trailing edge. Soars with wings in shallow V. Distinguished from Changeable by extensive barring on underparts, whitish-barred rump, and stronger dark barring on tail. Juvenile told from juvenile Changeable by more extensive dark streaking on crown and sides of head, white-tipped black crest, buff-barred rump, and fewer, more prominent tail-bars. Forested hills and mountains. Rare and irregular visitor to Neelum Valley and Murree Hills.

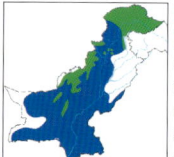

1 **COMMON KESTREL** *Falco tinnunculus* 32–35 cm
a b c MALE and **d e f** FEMALE Long, rather broad tail; wing-tips more rounded than on most falcons. Frequently hovers. Male has grey head and tail, and rufous upperparts heavily marked with black. Female and juvenile have rufous crown and nape streaked with black, diffuse and narrow dark moustachial stripe, rufous upperparts heavily marked with black, and dark barring on rufous tail. Common resident throughout Pakistan, supplemented by winter migrants from central Asia. Winter visitor throughout the plains, and breeds in both open forest and arid cliff areas throughout northern mountains up to Hunza. Avoids well-wooded and irrigated cultivated tracts.

2 **RED-NECKED FALCON** *Falco chicquera* 31–36 cm
a b ADULT and **c** JUVENILE Powerful falcon with pointed wings and longish tail. Flight usually fast and dashing. Has rufous crown and nape, pale blue-grey upperparts, white underparts finely barred with black, and grey tail with broad black subterminal band. Resident and largely sedentary throughout the Indus basin. Absent from Baluchistan and NWFP, except in foothill regions. Has declined drastically in recent decades due to the falconry trade.

3 **SOOTY FALCON** *Falco concolor* 33–36 cm
a b ADULT and **c d** JUVENILE Summer visitor. Slim, with very long wings and tail (the latter is wedge-shaped at tip). Adult entirely pale grey with blackish flight feathers; older males can be almost black. Juvenile has narrow buff fringes to upperparts, and yellowish-brown underparts and underwing-coverts, which are diffusely streaked. Adapted to arid coastal areas along main Palaearctic–Africa migration routes. A few pairs are summer breeders along the Mekran coast, especially in the newly established Hingol National Park.

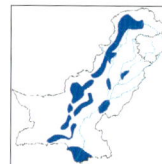

4 **MERLIN** *Falco columbarius* 25–30 cm
a b MALE and **c d** FEMALE Small and compact, with short, pointed wings. Flight typically swift, with rapid beats interspersed with short dashing glides, when wings often closed into body. Male has blue-grey upperparts, broad black subterminal tail-band, weak moustachial stripe, and diffuse patch of rufous orange on nape. Female and juvenile have weak moustachial stripe, brown upperparts with variable rufous/buff markings, and strongly barred uppertail. Rare winter visitor, mostly west of the Indus, including far northern areas. Hunts in mountain valleys, cropland and scrub desert.

FALCONS, PLATE 46

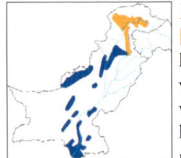

1 EURASIAN HOBBY *Falco subbuteo* 30–36 cm

a b ADULT and **c** JUVENILE Slim, with long, pointed wings and medium-length tail. Hunting flight is swift and powerful, with stiff beats interspersed with short glides. Adult has broad black moustachial stripe, cream underparts with bold blackish streaking, and rufous thighs and undertail-coverts. Juvenile has dark brown upperparts with buffish fringes, pale buffish underparts that are more heavily streaked, and lacks rufous thighs and undertail-coverts. Uncommon summer breeding visitor to northern areas. Uncommon winter visitor to foothills and valleys west of the Indus. During breeding season, adapted to outer Himalayan forest and northern alpine valleys. Up to 6 can be seen together in Murree Hills, hunting dragonflies.

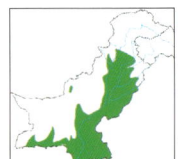

2 LAGGAR FALCON *Falco jugger* 43–46 cm

a b ADULT and **c d** JUVENILE Large falcon, although smaller, slimmer-winged and less powerful than Saker. Adult has rufous crown, fine but prominent dark moustachial stripe, dark brown upperparts, and rather uniform uppertail; underparts and underwing-coverts vary in extent of streaking, but lower flanks and thighs usually wholly dark brown; may show dark panel across underwing-coverts. Juvenile similar to adult, but crown duller and underparts very heavily streaked, and has greyish bare parts; differs from juvenile Peregrine in paler crown, finer moustachial stripe, and unbarred uppertail. Rare resident confined to more open, sparsely populated regions of the Indus basin. Absent from mountainous areas. Has suffered heavy persecution for falconry trade.

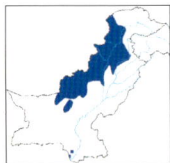

3 SAKER FALCON *Falco cherrug* 50–58 cm

a b c ADULT *F. c. cherrug* and **d** ADULT *F. c. milvipes* Large falcon with long wings and long tail. Wingbeats slow in level flight, with lazier flight action than Peregrine. At rest, tail extends noticeably beyond closed wings (wings fall just short of tail-tip on Laggar and are equal to tail on Peregrine). Adult *F. c. cherrug* has paler crown, less distinct moustachial stripe, and paler rufous-brown upperparts than Laggar; has less heavily marked underparts (with flanks and thighs usually clearly streaked and not appearing wholly brown, although some overlap exists). Outer tail feathers more prominently barred. Juvenile (not illustrated) has greyish cere and greyish legs and feet; otherwise similar to adult, but crown more heavily marked, moustachial stripe stronger, underparts more heavily streaked, and upperparts darker brown. *F. c. milvipes* has broad orange-buff barring on upperparts and is rather different in plumage from Laggar. Formerly frequent winter migrant visitor from Afghanistan and Tajikistan. Now almost extirpated, being most highly prized in falconry trade. Globally threatened (Endangered).

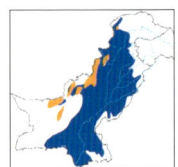

4 PEREGRINE FALCON *Falco peregrinus* 38–48 cm

a b ADULT and **c d** JUVENILE *F. p. peregrinator*, **e f** ADULT and **g h** JUVENILE *F. p. babylonicus*, and **i j** ADULT *F. p. calidus* Heavy-looking falcon with broad-based and pointed wings and short, broad-based tail. Flight strong, with stiff, shallow beats and occasional short glides. *F. p. calidus* has slate-grey upperparts, broad black moustachial stripe, and whitish underparts with narrow blackish barring; juvenile *calidus* (not illustrated) has browner upperparts, heavily streaked underparts, broad moustachial stripe, and barred uppertail. *F. p. peregrinator* has darker upperparts with more extensive black hood, and rufous underparts; juvenile *peregrinator* has darker brown upperparts than adult, and paler underparts with heavy streaking. *F. p. babylonicus* ('Barbary Falcon') has pale blue-grey upperparts, buffish underparts with only sparse markings, rufous on crown and nape, and finer moustachial stripe; juvenile *babylonicus* has darker brown upperparts, heavily streaked underparts, and only a trace of rufous on forehead and supercilium. *F. p. peregrinator* is an uncommon, locally migratory inhabitant of drier hilly tracts, where it occasionally breeds up to 2745 m. *F. p. babylonicus* is an uncommon, locally migratory species breeding in the drier hilly tracts of western Pakistan and wintering more widely in the Indus basin. *F. p. calidus* is a widespread winter visitor, mainly to the Indus basin, and is attracted especially to areas where congregations of waders feed. All forms have suffered massive declines in Pakistan due to the falconry trade.

GREBES, DARTER AND CORMORANTS, PLATE 47

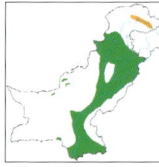

1 **LITTLE GREBE** *Tachybaptus ruficollis*   25–29 cm
a ADULT BREEDING and **b c** ADULT NON-BREEDING Small size, often with puffed-up rear end. In breeding plumage, has rufous cheeks and neck sides, and yellow patch at base of bill. In non-breeding plumage, has brownish-buff cheeks and flanks. Juvenile is similar to non-breeding but has brown stripes across cheeks. Whinnying calls confined to breeding season. Common widespread resident, mainly on open water bodies, occasionally on coastal creeks and main rivers.

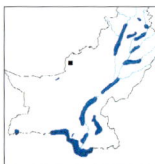

2 **GREAT CRESTED GREBE** *Podiceps cristatus*   46–51 cm
a ADULT BREEDING and **b c** ADULT NON-BREEDING Large and slender-necked. Has white cheeks and fore-neck in non-breeding plumage. Has rufous-orange ear-tufts and white cheeks and fore-neck in breeding plumage. (See Appendix for comparison with Red-necked Grebe.) Uncommon winter visitor, mainly to larger lakes, and more frequent in northern regions. Occasional breeding recorded on Khushdil Khan Lake, Baluchistan.

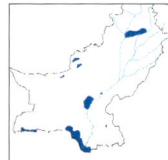

3 **BLACK-NECKED GREBE** *Podiceps nigricollis*   28–34 cm
a ADULT BREEDING and **b** ADULT NON-BREEDING Steep forehead, and bill appears upturned. Dusky ear-coverts contrast with white throat and sides of head in non-breeding plumage. Yellow ear-tufts, black neck and breast and rufous flanks in breeding plumage. (See Appendix for comparison with Horned Grebe.) Uncommon late winter visitor, with concentrations up to 200–300 on Salt Range and Baluchistan lakes. Occasional breeding recorded on both Zangi Nawar and Khushdil Khan lakes.

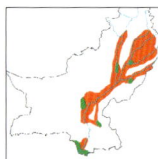

4 **DARTER** *Anhinga melanogaster*   85–97 cm
a b ADULT MALE BREEDING and **c** IMMATURE Long, slim head and neck, dagger-like bill, and long tail. Often swims with most of body submerged. In flight, neck is only partly outstretched, with kink at base. As with other cormorants, frequently perches with wings held outstretched to dry. Adult has white stripe down side of neck, lanceolate white scapular streaks, and white streaking on wing-coverts. Juvenile is buffish-white below, with buff fringes to coverts forming pale panel on upperwing. Widespread but uncommon resident on both lakes and main rivers, especially favouring barrage headponds. Breeds colonially in flooded trees.

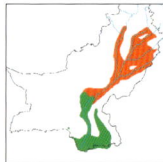

5 **LITTLE CORMORANT** *Phalacrocorax niger*   51 cm
a b ADULT BREEDING, **c** ADULT NON-BREEDING and **d** IMMATURE Small size with short bill. Adult breeding is all black, with a few white plumes on fore-crown and sides of head. Non-breeding is browner (and lacks white head plumes), with whitish chin. Immature has whitish chin and throat, and fore-neck and breast a shade paler than upperparts, with some pale fringes. (See Appendix for comparison with Pygmy Cormorant.) Abundant resident, more numerous in southern regions; up to 3000 counted on Lal Sohanran Lake. Breeds colonially with other cormorants and egrets in flooded trees. One colony *c.* 50 pairs in mangroves.

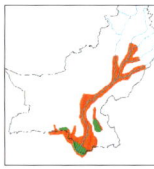

6 **INDIAN CORMORANT** *Phalacrocorax fuscicollis*   63 cm
a b ADULT BREEDING, **c** ADULT NON-BREEDING and **d** IMMATURE Smaller and slimmer than Great, with thinner neck, slimmer oval-shaped head, finer-looking bill, and proportionately longer tail. In flight, looks lighter, with thinner neck and quicker wing action. Larger than Little, with longer neck, oval-shaped head and longer bill. Adult breeding is glossy black, with tuft of white behind eye and scattering of white filoplumes on neck. Non-breeding lacks white plumes; has whitish throat, and browner-looking head, neck and underparts. Immature has brown upperparts and whitish underparts. Common resident, most numerous in southern regions. Better adapted to salt water than Little Cormorant. Breeds colonially in mangroves and flooded trees in fresh water.

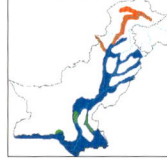

7 **GREAT CORMORANT** *Phalacrocorax carbo*   80–100 cm
a b ADULT BREEDING, **c** ADULT NON-BREEDING and **d** IMMATURE Large with thick neck and stout bill. Adult breeding is glossy black, with orange facial skin, white cheeks and throat, white head plumes and white thigh-patch. Non-breeding is more blackish brown, and lacks white head plumes and thigh patch. Immature has whitish or pale buff underparts. Partially resident and winter migrant, becoming abundant, with winter flocks *c.* 2000 roosting on delta sandbars. Less common on inland lakes and rivers.

EGRETS AND HERONS, PLATE 48

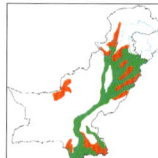

1 LITTLE EGRET *Egretta garzetta* 55–65 cm
a ADULT BREEDING and **b** ADULT NON-BREEDING Slim and graceful. Typically, has black bill, black legs with yellow feet, and greyish lores (lores reddish during courtship). Breeding plumage has prominent plumes on nape, breast and mantle. Abundant resident, especially Sind, breeding colonially with other waterbirds, northwards to Swat, but only occasional visitor to Baluchistan.

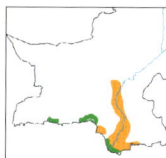

2 WESTERN REEF EGRET *Egretta gularis* 55–65 cm
a ADULT WHITE MORPH BREEDING, **b** ADULT DARK MORPH NON-BREEDING and **c** INTERMEDIATE MORPH Indus delta population occurs as 4 distinct colour morphs: all white 15%; slaty grey with small white wing-patch *c.* 70%; dark purplish grey with prominent white carpal wing-patch 10%; pale ashy grey with throat and lower belly white 5%. White morph can be difficult to separate from Little. Bill longer and stouter than Little's, and usually appearing very slightly downcurved. Legs also slightly shorter and thicker looking. Bill is usually mainly yellowish or brownish yellow, but may be black when breeding. Typically, has greenish or yellowish lores (although can be greyish as on Little, and Little can, exceptionally, appear to have yellowish tinge to lores). In breeding condition, the lores and base of bill are suffused purplish orange. Common resident along sea coast, especially in mangroves. Over 1000 counted during two-day survey on Indus delta. Nests colonially with other species.

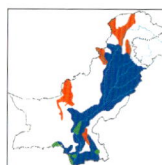

3 GREY HERON *Ardea cinerea* 90–98 cm
a b ADULT and **c** IMMATURE In flight, black flight feathers contrast with grey upperwing- and underwing-coverts. Adult has yellow bill, whitish head and neck with black head plumes, and black patches on belly. Immature has dark cap with variable crest, greyer neck, and lacks or has reduced black on belly. Common winter visitor from central Asia, with about 25% of population resident and breeding. Adapted to both marine and freshwater habitats.

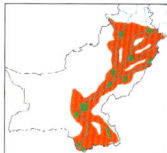

4 PURPLE HERON *Ardea purpurea* 78–90 cm
a b ADULT and **c d** JUVENILE Rakish, with long, thin neck. In flight, compared with Grey, bulge of recoiled neck is very pronounced, and protruding feet are large. Adult has chestnut head and neck with black stripes, grey mantle and upperwing-coverts, and dark chestnut belly and underwing-coverts. Juvenile has buffish neck and underparts, and brownish mantle and upperwing-coverts with rufous-buff fringes. Common resident, preferring reed-fringed lakes and rarely hunting in open exposed water. Nests often in single-species colonies.

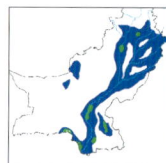

5 GREAT EGRET *Casmerodius albus* 65–72 cm
a ADULT BREEDING and **b** ADULT NON-BREEDING Large size, very long neck and large bill. Black line of gape extends behind eye. In breeding plumage, bill is black, lores blue and tibia reddish, and has prominent plumes on breast and mantle. In non-breeding plumage, bill is yellow and lores pale green. Common resident, preferring vicinity of larger lakes and main rivers. Nests colonially at onset of monsoon with other waterbirds.

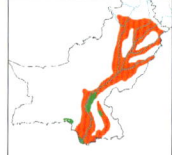

6 INTERMEDIATE EGRET *Mesophoyx intermedia* 65–72 cm
a ADULT BREEDING and **b** ADULT NON-BREEDING Smaller than Great, with shorter bill and neck. Black gape-line does not extend beyond eye. In breeding plumage, bill is black and lores are yellow-green, and it has breast- and mantle-plumes and a prominent crest. Has black-tipped yellow bill and yellow lores outside breeding season. Less common than Little or Great egrets. Widespread resident in Sind, especially East Nara, and Punjab. Nests colonially well after onset of monsoon.

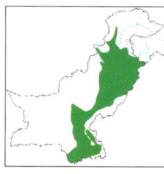

7 CATTLE EGRET *Bubulcus ibis* 48–53 cm
a b ADULT BREEDING and **c** ADULT NON-BREEDING Small, stocky egret, with short yellow bill and short legs. Has orange-buff on head, neck and mantle in breeding plumage. Common resident in Sind and Punjab up to foothills. Absent from Baluchistan and hilly areas of NWFP. Adapted to forage, especially with livestock, in drier grassland than other Ardeidae.

SMALL HERONS AND BITTERNS, PLATE 49

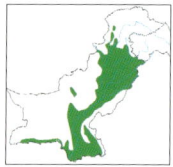

1 INDIAN POND HERON *Ardeola grayii*   42–45 cm
a b ADULT BREEDING and **c** ADULT NON-BREEDING Whitish wings contrast with dark saddle. Adult breeding plumage has yellowish-buff head and neck, white nape-plumes and maroon-brown mantle/scapulars. Head, neck and breast streaked in non-breeding plumage. Abundant resident throughout Sind, Punjab and NWFP, avoiding only mountainous areas. Forages in seepage pools, village tanks and mangroves.

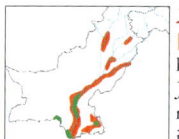

2 LITTLE HERON *Butorides striatus*   40–48 cm
a b ADULT and **c** JUVENILE Small, stocky and short-legged heron. Adult has black crown and long crest, dark greenish upperparts and greyish underparts. Juvenile has buff streaking on upperparts, and dark-streaked underparts. Uncommon resident, preferring perennial pools and well-vegetated channels. Irregular visitor to Punjab, absent from NWFP. Non-colonial nesting after onset of monsoon.

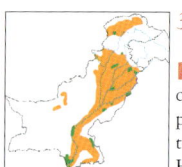

3 BLACK-CROWNED NIGHT HERON
Nycticorax nycticorax   58–65 cm
a b ADULT and **c** JUVENILE Stocky heron, with thick neck. Adult has black crown and mantle contrasting with grey wings and whitish underparts. Breeding plumage has elongated white nape-plumes. Juvenile is boldly streaked and spotted. Immature resembles juvenile but has unstreaked brown mantle/scapulars. Highly gregarious, common resident, breeding colonially. Some migrate to breed in northern areas with other waterbirds, high up in tall trees.

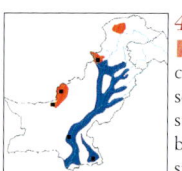

4 LITTLE BITTERN *Ixobrychus minutus*   33–38 cm
a b MALE, **c** FEMALE and **d** JUVENILE Small size. Buffish wing-coverts contrast with dark flight feathers in all plumages. Male has black crown and mantle/scapulars, and buff neck. Female has brown mantle/scapulars, with brownish-buff streaking on fore-neck. Juvenile has warm buff upperparts streaked with dark brown, and brown streaking on underparts; very similar to juvenile Yellow, but streaking on fore-neck and breast of Yellow is generally more rufous orange. Rare resident, breeding in lower Sind, Baluchistan and Kohat. Prefers tall reed-beds and permanent wetlands with submerged trees and bushes for nesting.

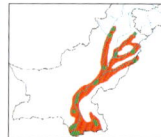

5 YELLOW BITTERN *Ixobrychus sinensis*   38 cm
a b MALE and **c** JUVENILE Small size. Yellowish-buff wing-coverts contrast with dark brown flight feathers. Male has pinkish-brown mantle/scapulars, and face and sides of neck are vinaceous. Female is similar to male but with rufous streaking on black crown, rufous-orange streaking on fore-neck and breast, and diffuse buff edges to rufous-brown mantle/scapulars. Juvenile appears buff with bold dark streaking to upperparts, including wing-coverts; fore-neck and breast are heavily streaked. Uncommon resident, preferring dense reed cover and perennial pools. Largely crepuscular in hunting. Unlike Little Bittern, nests in pure reed stands.

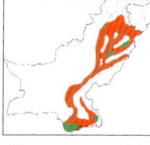

6 CINNAMON BITTERN *Ixobrychus cinnamomeus*   38 cm
a b MALE, **c** FEMALE and **d** JUVENILE Small size. Uniform-looking cinnamon-rufous flight feathers and tail in all plumages. Male has cinnamon-rufous crown, hind-neck and mantle/scapulars. Female has dark brown crown and mantle, and dark brown streaking on fore-neck and breast. Juvenile has buff mottling on dark brown upperparts, and is heavily streaked with dark brown on underparts. Uncommon resident, preferring extensive reed and sedge cover.

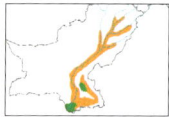

7 BLACK BITTERN *Dupetor flavicollis*   58 cm
a b ADULT and **c** JUVENILE Blackish upperparts, including wings, with orange-buff patch on side of neck. Juvenile has rufous fringes to upperparts. Uncommon resident, mainly confined to lower Sind and with population augmented by monsoon migrants. Prefers well-vegetated wetlands.

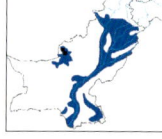

8 GREAT BITTERN *Botaurus stellaris*   70–80 cm
a b ADULT Stocky. Cryptically patterned with golden brown, blackish and buff. Uncommon winter migrant throughout main riverine tracts. Occasional breeding in Baluchistan. Prefers dense reed-beds.

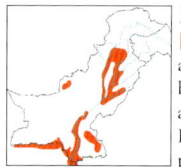

1 **GREATER FLAMINGO** *Phoenicopterus ruber*   125–145 cm
a b ADULT, **c** IMMATURE and **d** JUVENILE Larger than Lesser, with longer and thinner neck. Bill larger and less prominently kinked. Adult has pale pink bill with prominent dark tip, and variable amount of pinkish white on head, neck and body; in flight, crimson-pink upperwing-coverts contrast with whitish body. Immature has greyish-white head and neck, and white body lacking any pink; pink on bill develops with increasing age. Juvenile brownish grey, with white on coverts; bill grey, tipped with black, and legs grey. In flight, call is a goose-like honking. Forages in flocks in shallow brackish lakes and lagoons, uttering guttural gabbling. Irregular movement, sometimes in huge flocks. Most breed extralimitally, but breeding recorded in Badin District.

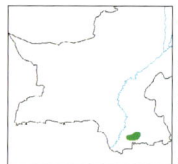

2 **LESSER FLAMINGO** *Phoenicopterus minor*   80–90 cm
a b ADULT, **c** IMMATURE and **d** JUVENILE Smaller than Greater Flamingo; neck appears shorter, and bill is smaller and more prominently kinked. Adult has black-tipped dark red bill, dark red iris and facial skin, and deep rose-pink on head, neck and body; blood-red centres to lesser and median upperwing-coverts contrast with paler pink of rest of coverts. Immature has greyish-brown head and neck, pale pink body and mainly pink coverts; bill coloration develops with increasing age. Juvenile mainly grey-brown, with dark-tipped purplish-brown bill and grey legs. Much less common then Greater, but over 2000 recorded breeding in Badin District.

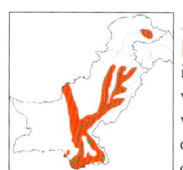

3 **GLOSSY IBIS** *Plegadis falcinellus*   55–65 cm
a ADULT BREEDING, **b** ADULT NON-BREEDING and **c** JUVENILE Small, dark ibis with rather fine down-curved bill. Adult breeding is deep chestnut, glossed with purple and green, and with metallic green-and-purple wings; has narrow white surround to bare lores. Adult non-breeding is duller, with white streaking on dark brown head and neck. Juvenile is similar to adult non-breeding, but is dark brown with white mottling on head, and has only faint greenish gloss to upperparts. Common only in lower Sind, with resident population augmented by winter migrants from central Asia and Iran. Prefers flooded paddies and freshwater lake margins.

4 **BLACK-HEADED IBIS** *Threskiornis melanocephalus*   75 cm
a ADULT BREEDING and **b** IMMATURE Stocky, mainly white ibis with stout down-curved black bill. Adult breeding has naked black head, white lower-neck plumes, variable yellow wash to mantle and breast, and grey on scapulars and elongated tertials. Adult non-breeding has all-white body and lacks neck plumes. Immature has grey feathering on head and neck, and black-tipped wings. Rare and largely sedentary. Confined to lower Sind, especially tidal creeks, mangroves and seepage zones around larger lakes.

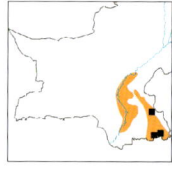

5 **BLACK IBIS** *Pseudibis papillosa*   68 cm
ADULT Stocky, dark ibis with relatively stout, down-curved bill. Has white shoulder-patch and reddish legs. Adult has naked black head with red patch on rear crown and nape, and is dark brown with green-and-purple gloss. Immature is dark brown, including feathered head. Rare; mainly a non-breeding monsoon-season visitor to freshwater wetlands in East Nara, Indus delta and lower Sind. Often forages in dry cultivation.

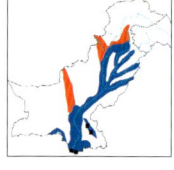

6 **EURASIAN SPOONBILL** *Platalea leucorodia*   80–90 cm
a ADULT BREEDING and **b** JUVENILE White, with spatulate-tipped bill. Adult has black bill with yellow tip; has crest and yellow breast-patch when breeding. Juvenile has pink bill; in flight, shows black tips to primaries. Frequent winter migrant, with about one-third of population consisting of breeding residents. Prefers large lakes and rivers where it can forage in shallow water.

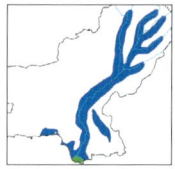

1 PAINTED STORK *Mycteria leucocephala*   93–100 cm

a b ADULT and **c d** IMMATURE Adult has down-curved yellow bill, bare orange-yellow or red face, and red legs; white barring on mainly black upperwing-coverts, pinkish tertials, and black barring across breast. Juvenile is dirty greyish white, with grey-brown (feathered) head and neck, and brown lesser coverts; bill and legs duller than adult's. Rare resident; has suffered persecution in nesting colonies in Indus mangroves by fishermen selling chicks to animal exporters. Roosts gregariously in trees; daytime foraging in vicinity of main rivers, preferring shallow water.

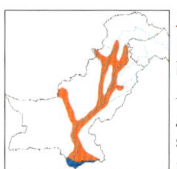

2 WHITE STORK *Ciconia ciconia*   100–125 cm

a b ADULT Mainly white stork, with black flight feathers and striking red bill and legs. Generally has cleaner black-and-white appearance compared with Asian Openbill (see Appendix). Juvenile is similar to adult but with brown greater coverts and duller brownish-red bill and legs. Red bill and white tail help separate it from Asian Openbill at a distance in flight. Rare winter migrant, recorded on passage throughout major river systems and across northern Baluchistan. A small population overwinters in southern Sind, the majority continuing southwestwards into India.

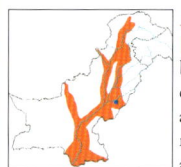

3 BLACK STORK *Ciconia nigra*   90–100 cm

a b ADULT and **c** IMMATURE Adult mainly glossy black, with white lower breast and belly, and red bill and legs; in flight, white underparts and axillaries contrast strongly with black neck and underwing. Juvenile has brown head, neck and upperparts flecked with white; bill and legs greyish green. Scarce winter migrant, with small flocks observed travelling along river valleys in northern areas, and across northern Baluchistan. Adapted to forage in drier areas than White Stork, with numbers entering Pakistan on spring passage from Rajasthan. Flocks observed on Khushdil Khan Lake, Baluchistan, as late as 21 June.

4 BLACK-NECKED STORK *Ephippiorhynchus asiaticus*   129–150 cm

a b ADULT and **c d** IMMATURE Large black-and-white stork with long red legs and huge black bill. In flight, wings are white except for broad black band across coverts, and tail is black. Male has brown iris; yellow in female. Juvenile has fawn-brown head, neck and mantle, mainly brown wing-coverts and mainly blackish-brown flight feathers; legs dark. Very rare irregular visitor to major lakes in both Sind and Punjab. Formerly bred in mangroves in Indus delta but not recorded since late 1970s. Prefers to hunt in shallow water, taking fish, swimming reptiles and unwary waterbirds.

5 WOOLLY-NECKED STORK *Ciconia episcopus*   75–92 cm

a b ADULT Stocky, largely blackish stork with 'woolly' white neck, black 'skullcap', and white vent and undertail-coverts. In flight, upperwing and underwing are entirely dark. Juvenile is similar to adult but with duller brown body and wings, and feathered forehead. There are breeding records from the 20th century. It may still survive in Daphar irrigated forest plantation, and there are recent unconfirmed reports from along the Ravi river, adjacent to the Indian border in Kasur District.

PELICANS AND MISCELLANEOUS SEABIRDS, PLATE 52

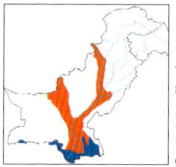

1 GREAT WHITE PELICAN *Pelecanus onocrotalus*   140–175 cm
a ADULT BREEDING, **b c** ADULT NON-BREEDING, **d** IMMATURE and **e** JUVENILE Adult and immature have black underside to primaries and secondaries, which contrast strongly with white (or largely white) underwing-coverts. Feathering of forehead narrower than on Dalmatian, and tapers to a point at bill-base (as orbital skin more extensive). Adult breeding has white body and wing-coverts tinged with pink, bright yellow pouch and pinkish skin around eye. Adult non-breeding has duller bare parts and lacks pink tinge and white crest. Immature has variable amounts of brown on wing-coverts and scapulars. Juvenile has largely brown head, neck and upperparts, including upperwing-coverts, and brown flight feathers; upperwing appears more uniform brown, and underwing shows pale central panel contrasting with dark inner coverts and flight feathers; greyish pouch becomes yellower with age. Uncommon winter visitor, mainly to Thatta and Badin districts; sympatric with Dalmatian. Graceful circling on thermals but clumsy when slapping feet synchronously to take off from sandbars.

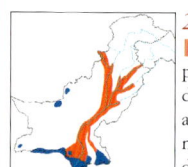

2 DALMATIAN PELICAN *Pelecanus crispus*   160–180 cm
a ADULT BREEDING, **b c** ADULT NON-BREEDING and **d** IMMATURE In all plumages, has greyish underside to secondaries and inner primaries (becoming darker on outer primaries), lacking strong contrast with pale underwing-coverts, and often with whiter central panel. Forehead feathering broader across upper mandible (orbital skin more restricted than on Great White). Legs and feet always dark grey (pinkish on Great White). Adult breeding has orange pouch and purple skin around eye, and curly or bushy crest. Adult non-breeding more dirty white; pouch and skin around eye paler. Immature dingier than adult non-breeding, with some pale grey-brown on upperwing-coverts and scapulars. Juvenile has pale grey-brown mottling on hind-neck and upperparts, including upperwing-coverts; pouch greyish yellow. Uncommon winter visitor, formerly more numerous than Great White. Both heavily persecuted on migration for supposed medicinal properties. Globally threatened (Vulnerable).

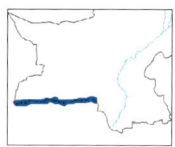

3 RED-BILLED TROPICBIRD *Phaethon aethereus*   48 cm
a b ADULT and **c** JUVENILE Adult has red bill, white tail-streamers, black barring on mantle and scapulars, and much black on primaries. Juvenile has yellow bill with black tip, and black band across nape. Rare visitor, mainly sub-adult birds along western coast of Baluchistan from January to May.

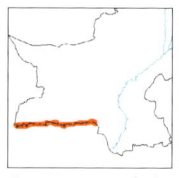

4 MASKED BOOBY *Sula dactylatra*   81–92 cm
a ADULT, **b** IMMATURE and **c** JUVENILE Large and robust booby. Adult largely white, with black mask and black flight feathers and tail. Juvenile has brown head, neck and upperparts, with whitish collar and whitish scaling on upperparts; underbody white, and shows much white on underwing-coverts. Head, upper body and upperwing-coverts of immature become increasingly white with age. Rare and mainly sub-adult birds regularly recorded off Makran Coast, hunting up to 48 km offshore.

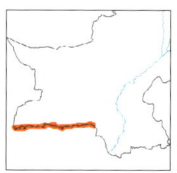

5 PERSIAN SHEARWATER *Puffinus persicus*   30.5–33 cm
a b ADULT A small shearwater with brown upperparts and whitish underparts. Slightly larger than the similar Audubon's Shearwater (which is recorded from the Indian subcontinent, although not from Pakistan). Separated from that species by browner coloration to upperparts, longer bill, less extensive white on underwing, and brownish axillaries and flanks. Hunting flocks up to 200 birds regularly sighted, usually well offshore along Arabian Sea coast, mainly post-monsoon and during the winter months.

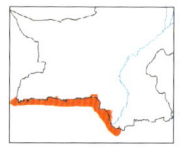

6 WILSON'S STORM-PETREL *Oceanites oceanicus*   15–19 cm
a b ADULT Dark underparts and underwing. Small, mainly dark petrel with white rump-band. Feet project noticeably beyond tail. When fluttering over water, dangling feet show yellow webbing. Though post-monsoon coastal upwelling attracts this surface feeder well inshore, non-breeders are recorded in all months of the year, even hovering over Green Turtles and human bathers.

1 INDIAN PITTA *Pitta brachyura*   19 cm
ADULT Bold black stripe through eye, white throat and supercilium, buff lateral crown-stripes, and buff breast and flanks. Calling males utter double-noted whistle, *wheet-tieear*. Scarce and local summer breeding visitor from May to October. Haunts thickest scrub-covered ravines in Margalla Hills.

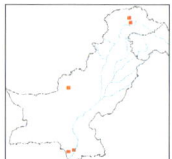

2 RED-BACKED SHRIKE *Lanius collurio*   17 cm
a MALE, **b** FEMALE and **c** 1ST-WINTER Compared with Bay-backed, male lacks broad black forehead, has rufous mantle, and lacks white wing-patch. Female has grey cast to head and nape, and dark brown tail. First-winter similar to female, but has barred upperparts. Scarce passage migrant in autumn only, September to late October, along the Indus. Prefers scrub-covered, uncultivated tracts.

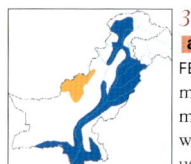

3 RUFOUS-TAILED SHRIKE *Lanius isabellinus*   18–19 cm
a MALE, **b** FEMALE and **c** 1ST-WINTER *L. i. phoenicuroides*, and **d** MALE, **e** FEMALE and **f** 1ST-WINTER *L. i. isabellinus* Has pale sandy-brown/grey-brown mantle and warm rufous rump and tail. Male has small white patch at base of primaries, which is lacking in female Red-backed. Female is similar to male but lacks white patch on wing and has grey-brown (rather than blackish) ear-coverts, and usually has some scaling on underparts. First-winter birds are similar to female but with pale fringes and dark subterminal lines to scapulars, wing-coverts and tertials. *L. i. phoenicuroides* has rufous cast to crown and nape, contrasting with grey-brown mantle. Uncommon winter migrant throughout major riverine tracts, from far north to sea coast. Favours scrub forest with tamarisk, avoiding cultivated tracts. A small population breeds on bush-studded slopes and ravines in Baluchistan.

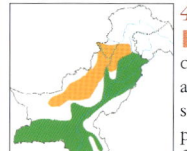

4 BAY-BACKED SHRIKE *Lanius vittatus*   17 cm
a ADULT, **b** IMMATURE and **c** JUVENILE Adult has black forehead and mask contrasting with pale grey crown and nape, deep maroon mantle, whitish rump and white patch at base of primaries. Juvenile told from juvenile Long-tailed by smaller size and shorter tail, more uniform greyish/buffish base colour to upperparts, pale rump, more intricately patterned wing-coverts and tertials (with buff fringes and dark subterminal crescents and central marks), and primary coverts that are prominently tipped with buff. First-year like washed-out version of adult; lacks black forehead. (See Appendix for description of similar Woodchat Shrike.) Breeding males call *chew-chew*, also mimetic phrases of other birds. Common resident of southern Baluchistan, and throughout Sind and Punjab, except desert areas. Summer visitor breeding in northern Baluchistan and NWFP up to 1300 m.

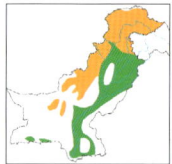

5 LONG-TAILED SHRIKE *Lanius schach*   25 cm
a ADULT and **b** JUVENILE Adult has grey mantle, rufous scapulars and upper back, narrow black forehead, rufous sides to black tail, and small white patch on primaries. Juvenile has (dark-barred) rufous-brown scapulars, back and rump; dark greater coverts and tertials fringed rufous. Song includes mimicry of other birds, interspersed with harsh *kech-kech* calls. Very common resident, sympatric with *L. vittatus*, with summer breeders migrating into far northern areas, up to 3000 m.

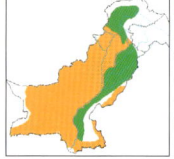

6 SOUTHERN GREY SHRIKE *Lanius meridionalis*   24 cm
a ADULT and **b** JUVENILE Narrow black forehead and broad black mask, grey mantle with white scalulars, broad white tips to secondaries, white sides and tip to tail, and white underparts. Juvenile has sandy cast to grey upperparts, buff tips to tertials and coverts, and grey mask. (See Appendix for comparison with Great Grey Shrike.) Adapted to more open arid tracts than other shrikes. A common summer breeding visitor and resident over much of Pakistan, avoiding only forested areas.

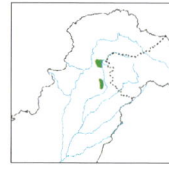
1 EURASIAN JAY *Garrulus glandarius* 32–36 cm
ADULT Reddish-brown head and body, black moustachial stripe, and blue barring on wings. White rump contrasts with black tail in flight. Though shy and wary, harsh warning calls betray its presence. Uncommon resident in forested Himalayan areas with a good admixture of deciduous trees, especially *Quercus* spp., 900–2750 m. Often in company with Black-headed Jays or Yellow-billed Blue Magpies.

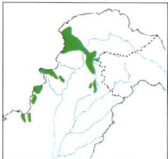
2 BLACK-HEADED JAY *Garrulus lanceolatus* 33 cm
ADULT Black face and crest, streaked throat, and pinkish-fawn body; blue barring on wings and tail. Common resident throughout Himalayan forest, and better adapted to drier, lower elevations in Baluchistan and NWFP, and colder northern forest than Eurasian Jay. Descends to main valleys at 1200 m in winter.

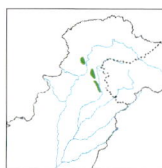
3 YELLOW-BILLED BLUE MAGPIE
Urocissa flavirostris 61–66 cm
ADULT Yellow bill, white crescent on nape, blue upperparts, and very long, graduated, black-and-white-tipped tail. Juvenile has olive-yellow bill. Warning calls are a rapidly descending *chik-kwitch-kitch-tich-ich*. Frequent resident of Himalayan moist temperate forest, extending westwards into Swat and Indus Kohistan, between 900–3350 m. Descends to lower forested slopes in winter, but never as far south as Margalla Hills. Always forages in small parties and is omnivorous in diet.

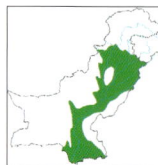
4 RUFOUS TREEPIE *Dendrocitta vagabunda* 46–50 cm
ADULT Slate-grey hood, rufous-brown mantle, pale grey wing-panel, buffish underparts and rump, and whitish subterminal tail-band. Juvenile has brown hood. Its harsh alarm cries, *ka-kak-kaakh-kaakh*, warn other birds and mammals of potential threats. Common resident, widespread throughout Indus plains and foothills up to 2300 m. Prefers better tree or bush cover. Largely arboreal and single when hunting.

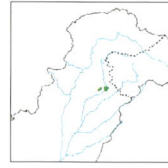
5 GREY TREEPIE *Dendrocitta formosae* 36–40 cm
ADULT Dark grey face, grey underparts and rump, and black wings with white patch at base of primaries. Juvenile duller version of adult. Rare and local, largely confined to Margalla Hills and outer foothills northeast of Islamabad, between 460 m and 1060 m, where it is often sympatric with Rufous Treepie. Not as vocal as the former and more secretive. Omnivorous in diet, taking nestling birds, berries and ground-dwelling insects.

6 BLACK-BILLED MAGPIE *Pica pica* 43–50 cm
ADULT Black and white, with long metallic green/purple tail. Usually seen in pairs year-round. Both sexes frequently utter loud chattering calls, *shak-shak-shak*, indicating either excitement, threat or alarm. Typically corvine, combining wariness with inquisitiveness. Omnivorous in diet. Common, though thinly distributed throughout drier mountainous areas, from Baluchistan to the far northern borders.

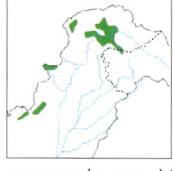
7 SPOTTED NUTCRACKER *Nucifraga caryocatactes* 32–35 cm
ADULT Mainly brown, with heavy white spotting on head and body. In flight, shows white tips to wing feathers, and white sides and tip to tail. Very vocal, uttering territorial harsh *kraak* calls, with longer twanging/wheezing songs in breeding season. Locally frequent resident, mainly in Himalayan cold, dry forest, from Chitral east to Baltistan, with isolated populations in Safed Koh, Shingar and south Waziristan. Specialist feeder on coniferous seeds, which it harvests and stores underground in seasons of abundance.

CHOUGHS AND CROWS, PLATE 55

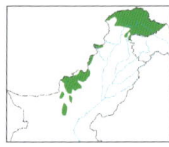

1 RED-BILLED CHOUGH *Pyrrhocorax pyrrhocorax* 36–40 cm
ADULT Curved red bill (shorter and orange-brown on juvenile). Call is a far-carrying, penetrating and nasal *chaow…chaow*. Abundant resident in far northern areas, and less common in northern Baluchistan. Better adapted to drier mountainous regions than Yellow-billed, though often sympatric with latter. Winters gregariously in valleys, summer breeding up to 5000 m.

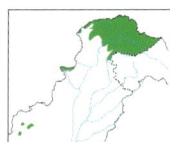

2 YELLOW-BILLED CHOUGH *Pyrrhocorax graculus* 37–39 cm
ADULT Almost straight yellow bill (olive-yellow on juvenile). Call is a far-carrying, rippling *preeep* and a descending whistled *sweeeoo*. Habits are similar to Red-billed, but generally occurs at higher altitude. Common resident above 3660 m, throughout northern mountainous areas, with small isolated population confined to highest ridges in Baluchistan and NWFP. Winter flocks feed with Red-billed, but better adapted to forage in vicinity of human settlements, and to endure severe snowstorms.

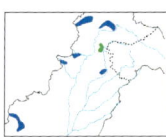

3 EURASIAN JACKDAW *Corvus monedula* 34–39 cm
ADULT Small size. Grey nape and hind-neck. Adult has pale grey iris. Isolated local resident population 1200–1500 m in Hazara District between Mansehra and Balakot, with winter migrant flocks in main valleys in northern areas, Kohat, Rawalpindi and Quetta districts.

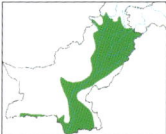

4 HOUSE CROW *Corvus splendens* 40 cm
ADULT Two-toned appearance, with paler nape, neck and breast. Very abundant commensal species, occurring throughout Indus basin, especially around human habitation. Absent from pure desert and dry mountainous areas above 600 m.

5 ROOK *Corvus frugilegus* 47 cm
a **b** ADULT and **c** JUVENILE Long, pointed bill, steep forehead, and baggy 'trousers'. Adult has bare white skin at base of bill and on throat. Erratic winter visitor to cultivation and pastures, in flocks from main valleys of Baltistan to Chitral and the west bank of the Indus, south to Quetta District.

6 CARRION CROW *Corvus corone* 48–56 cm
a **b** ADULT *C. c. orientalis* and **c** ADULT *C. c. sharpii* Resident (*C. c. orientalis*) and winter visitor (*C. c. sharpii*). Comparatively straight bill and flat crown. Race *sharpii* is two-toned like House Crow, but has black head and breast and grey mantle. *C. c. orientalis* is a scarce resident and partial winter migrant in main valleys in Baltistan, Gilgit and upper Kurram. *C. c. sharpii* is a more frequent winter visitor only to Baltistan, Gilgit and western parts of NWFP. Open country and cultivation, 2700–3600 m.

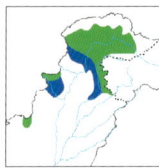

7 LARGE-BILLED CROW *Corvus macrorhynchos* 46–59 cm
a **b** ADULT All black, lacking paler collar of House Crow. Domed head, and large bill with arched culmen. Best told from Raven by absence of throat hackles, shorter and broader wings, less strongly wedge-shaped tail, squarer or domed crown, and dry *kaaa-kaaa* call. Common resident throughout northern areas, adapted to coniferous forest and open alpine areas, up to 3660 m. Isolated populations in Safed Koh and Shingar. Winters down to Islamabad. Forest, cultivation and open country above the tree-line; usually associated with towns and villages; up to 4500 m.

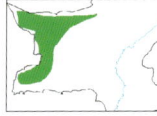

8 BROWN-NECKED RAVEN *Corvus ruficollis* 52–56 cm
a **b** ADULT Brown cast to head, neck and breast; lacks prominent throat hackles. Slimmer than Common Raven, with different call. Frequent resident in desert regions of southwest Baluchistan. Often in flocks, attracted to refuse tips and small cultivated fields.

9 COMMON RAVEN *Corvus corax* 58–69 cm
a **b** ADULT Very large; long and angular wings, prominent throat hackles, and wedge-shaped tail. Call is a loud, deep, resonant, croaking *wock…wock*, different from other crows. Thinly distributed resident confined to more arid uncultivated tracts throughout the Indus basin, Baluchistan and far northern mountains.

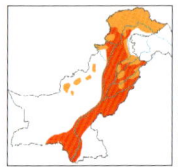

1 **EURASIAN GOLDEN ORIOLE** *Oriolus oriolus*   25 cm

a MALE, **b** FEMALE and **c** IMMATURE Male golden yellow, with black mask and mainly black wings. Female and immature variable, usually with streaking on underparts and yellowish-green upperparts. Nominate subspecies is a double passage migrant from East Africa to central Asia. Indian subspecies *O. o. kundoo* is a summer visitor, breeding in northern mountain valleys and sporadically in northern Punjab. Song is a loud, fluty *weela-wheooh*.

2 **LARGE CUCKOOSHRIKE** *Coracina macei*   30 cm

a MALE and **b** FEMALE Large and mainly pale grey in coloration. Female has grey barring on underparts. Song is a rich, fluty *pi-io-io*. Rare straggler to outer foothills of Murree Hills eastwards. Formerly considered an uncommon resident.

3 **BLACK-WINGED CUCKOOSHRIKE**
    Coracina melaschistos   24 cm

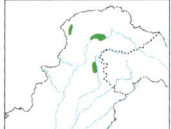

a MALE and **b** FEMALE Male slate-grey, with black wings and bold white tips to tail feathers. Female paler grey, with faint barring on underparts. Uncommon summer breeding visitor to Murree Hills, usually below 1800 m, preferring predominantly deciduous forest. Not recorded further west. Winters extralimitally.

4 **ROSY MINIVET** *Pericrocotus roseus*   20 cm

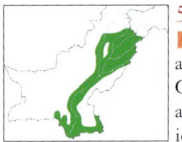

a MALE and **b** FEMALE Male has grey-brown upperparts, white throat, and pinkish underparts and rump. Female has greyish forehead, white throat, pale yellow underparts, and dull olive-yellow rump. Hunts in parties, while constantly calling *tiss-sit-it it*. Rare resident in subtropical Pine zone of drier outer forested slopes from Murree Hills to Hazara District, 1200–1800 m.

5 **SMALL MINIVET** *Pericrocotus cinnamomeus*   16 cm

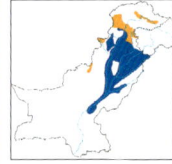

a MALE and **b** FEMALE Small size. Male has grey upperparts, dark grey throat, and orange on underparts. Female has pale throat and orange wash on underparts. Common resident below 1500 m in Sind, Punjab and NWFP, avoiding cropland and confined to irrigated plantations, relict patches of riverine forest and subtropical thorn scrub.

6 **LONG-TAILED MINIVET** *Pericrocotus ethologus*   20 cm

a MALE and **b** FEMALE Male is the only 'red-and-black' minivet occurring in Pakistan. Extension of red wing-patch down secondaries is best feature separating it from similar species found elsewhere in the subcontinent. Female has narrow, indistinct yellow wash on forehead and supercilium, grey ear-coverts, and pale yellow throat becoming more orange-yellow on breast (compare with female Rosy). Common and locally migratory resident. Breeds in Himalayan subtropical, cool temperate and inner drier forest. Winters on adjacent plains. Hunting parties constantly utter loud interrogative whistles, *tsweeh-tsweeh twooh*.

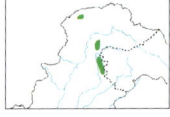

7 **YELLOW-BELLIED FANTAIL** *Rhipidura hypoxantha*   13 cm

MALE Long, fanned tail, yellow supercilium, dark mask and yellow underparts. Forests and high-altitude shrubberies. Rare winter visitor to outer hills and adjacent plains, from Sialkot west to Margalla Hills.

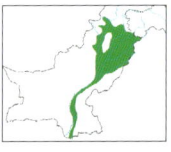

8 **WHITE-THROATED FANTAIL** *Rhipidura albicollis*   19 cm

ADULT Narrow white supercilium and white throat; lacks spotting on wing-coverts. Scarce resident, confined to Margallas and outer foothills of Murree Hills, and eastwards to Kahuta. Prefers damper, shadier ravines. Song consists of 4 or 5 spaced human-like whistles.

9 **WHITE-BROWED FANTAIL** *Rhipidura aureola*   18 cm

ADULT Broad white supercilia, which meet over forehead, blackish throat, white breast and belly, and white spotting on wing-coverts. Common sedentary resident throughout Punjab, but confined to riverine forest in Sind. Adapted to tree-lined canals and old bungalow gardens. Characteristic restless posturing when perched, with jerky flight.

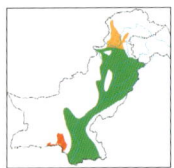

1 **BLACK DRONGO** *Dicrurus macrocercus*   28 cm
a ADULT and **b** IMMATURE Adult has glossy blue-black underparts and white rictal spot. Tail-fork may be lost during moult. Immature has black underparts with bold whitish fringes. Both sexes indulge in noisy two-noted territorial calls, *chee-ooh*. Common resident of Sind and Punjab, and summer migrant to main valleys in northern areas, up to 2600 m. Prefers hunting in open country from high observation post, especially telephone wires.

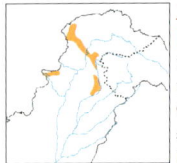

2 **ASHY DRONGO** *Dicrurus leucophaeus*   29 cm
a ADULT and **b** IMMATURE Adult has dark grey underparts and slate-grey upperparts with blue-grey gloss; iris bright red. Immature has brownish-grey underparts with indistinct pale fringes. Frequent summer breeding visitor from Chitral eastwards, 900–2500 m, in temperate Himalayan forest. Prefers lower altitudes with good admixture of deciduous trees. Can be sympatric with Black Drongo in Murree Hills. Like its congener, it is particularly noisy at dawn and dusk, preferring to hunt from the top of tall trees.

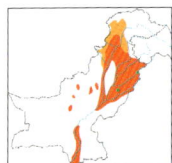

3 **ASIAN PARADISE-FLYCATCHER**
Terpsiphone paradisi   20 cm
a WHITE MALE, **b** RUFOUS MALE and **c** FEMALE Male has black head and crest, with white or rufous upperparts and long tail-streamers. Female and immatures have reduced crest and lack streamers. Frequently utters harsh advertising calls, *quieetch*. Nesting males have a soft yodelling song of 5 or 6 whistles. Locally common summer visitor, recorded on passage from Indus delta to Himalayas. Occasionally breeds in northern Punjab in shady well-watered locations, and regularly breeds in main Himalayan valleys up to 1500 m, Kurram, and Chitral east to Murree foothills.

4 **COMMON WOODSHRIKE** *Tephrodornis pondicerianus*   18 cm
ADULT Dark mask and broad white supercilium. Has white sides to tail. Strictly arboreal in hunting. Interrogative piping whistles betray its presence. Common sedentary resident throughout Indus basin, favouring larger tree-lined canals, irrigated forest plantations, riverine forest and uncultivated areas with subtropical thorn scrub cover.

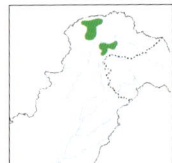

5 **WHITE-THROATED DIPPER** *Cinclus cinclus*   20 cm
a ADULT and **b** JUVENILE Adult has white throat and breast. Juvenile has dark scaling on grey upperparts, and grey scaling on whitish underparts. Very early breeder, when male gives rapid wren-like warbling song, audible above hissing water. Scarce mountain dweller, favouring alpine streams above 3000–4250 m, even when partly frozen over in winter. The Brown Dipper often occurs lower down on the same streams.

6 **BROWN DIPPER** *Cinclus pallasii*   20 cm
a ADULT and **b** JUVENILE Adult entirely brown. Juvenile has pale spotting on brown upperparts and underparts. Common resident on all Himalayan streams and smaller rivers, often occurring sympatrically with White-throated Dipper, up to 3900 m. A few (juveniles?) may descend to outer foothills in winter, but majority remain on their territories even when streams are partly frozen over.

THRUSHES, PLATE 58

1 RUFOUS-TAILED ROCK THRUSH *Monticola saxatilis* 20 cm
a MALE BREEDING, **b** FEMALE and **c** 1ST-WINTER MALE Distinguished from other species of rock thrush by orange-red uppertail-coverts and tail (in all plumages). Male has bluish head/mantle, white back and orange underparts, which are obscured by pale fringes in non-breeding and first-winter plumages. Female has scaling on upperparts, and orange wash to scaled underparts. Uncommon, with marked spring passage through drier western hilly tracts, and occasional breeding in northern Baluchistan.

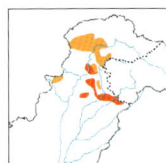

2 BLUE-CAPPED ROCK THRUSH
Monticola cinclorhynchus 17 cm
a MALE BREEDING, **b** FEMALE and **c** 1ST-WINTER MALE Male has white wing-patch, blue crown and throat, and orange rump and underparts; bright coloration obscured by pale fringes in non-breeding and first-winter plumages. Female has uniform olive-brown upperparts, including tail; lacks buff neck-patch of Chestnut-bellied. Song is a short sequence of rising and falling phrases. Very common summer breeding visitor throughout Himalayan forested regions, from 900 m in subtropical Pine to 3000 m in spruce and birch. Prefers drier, more exposed slopes than the moister shadier areas favoured by the Chestnut Thrush. Winters in South India.

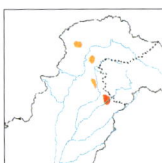

3 CHESTNUT-BELLIED ROCK THRUSH
Monticola rufiventris 23 cm
a MALE and **b** FEMALE Male has chestnut-red underparts and blue rump and uppertail-coverts; lacks white on wing. Female has orange-buff neck-patch, dark barring on slaty olive-brown upperparts, and heavy scaling on underparts. Males sing from treetops, a rather weak warbling of 5 or 6 up and down whistles. Locally frequent resident, breeding at 2000–3000 m in Himalayan moist temperate forest. It favours precipitous rocky terrain for nesting (on the ground).

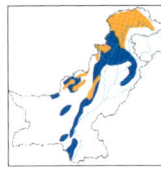

4 BLUE ROCK THRUSH *Monticola solitarius* 20 cm
a MALE BREEDING, **b** FEMALE and **c** 1ST-WINTER MALE Male indigo-blue, with bright coloration obscured by pale fringes in non-breeding and first-winter plumages. Female has bluish cast to slaty-brown upperparts, and buff scaling on underparts. Breeding males sing from prominent rocks and during flight display; a brief melodious sequence of elided warbles. The commonest rock thrush in Pakistan, adapted to drier mountain tracts, and breeding widely in northern Baluchistan and the far north from Chitral eastwards to Baltistan, 1500–4500 m. Part of the population migrates to winter in peninsular India, but the majority winter in the drier foothills on the west bank of the Indus and the Punjab Salt Range.

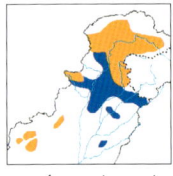

5 BLUE WHISTLING THRUSH *Myophonus caeruleus* 33 cm
ADULT Adult blackish, spangled with glistening blue; bill yellow. Juvenile browner, and lacks blue spangling. Breeding males sing pre-dawn, with short piercing whistles, and year-round have a more grating jay-like territorial song. Common resident, breeding throughout northern areas and in shady damper ravines in Baluchistan. Always occurs in proximity to water, from large rivers to tiny non-perennial rivulets. Many remain year-round in northern mountains, but numbers migrate in winter to adjacent plains.

6 ORANGE-HEADED THRUSH *Zoothera citrina* 21 cm
a MALE and **b** FEMALE Adult has orange head and underparts; male with blue-grey mantle, female with olive-brown wash to mantle. Males sing from May to August, repeated melodious warbling strophes, reminiscent of song of Mistle Thrush. Rare and local summer breeder to outer Himalayan foothills. Confined to dense scrub jungle in vicinity of streams. Not recorded west of Margallas. Winter migrant to India.

1 **PLAIN-BACKED THRUSH** *Zoothera mollissima*   27 cm

ADULT Uniform upperparts is the best feature separating it from Scaly Thrush, and barred underparts the best feature separating it from Mistle Thrush. The very similar Long-tailed Thrush is not known from Pakistan. The main features distinguishing it from that species are absent or indistinct wing-bars, more rufescent coloration to upperparts, less pronounced pale wing-panel, more extensive black scaling on belly and flanks, and shorter-looking tail. Song is similar to that of Scaly Thrush, consisting of spaced chirps and whistles. Alarm call is a grating rattle. Very rare high-altitude resident, confined to moist alpine slopes, 3600–4250 m, from Hazara District east to Azad Kashmir. Limited altitudinal movement to forested areas in winter.

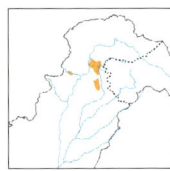

2 **SCALY THRUSH** *Zoothera dauma*   26–27 cm

ADULT Boldly scaled upperparts and underparts, and pale panels in wing. Juvenile has spotted breast. Unusual song, consisting of well-spaced phrases uttered in combination for up to 15 minutes, along with mournful whistles, chirrups and tremulo warbles. Scarce summer breeder in Himalayan moist coniferous forest with good understorey of shrubs, 2100–3000 m. Winters in adjacent plains but mostly eastwards along Siwaliks, into India.

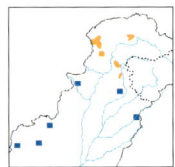

3 **TICKELL'S THRUSH** *Turdus unicolor*   21 cm

a MALE, **b** FEMALE and **c** 1ST-WINTER MALE Small thrush. Male is pale bluish grey, with whitish belly and vent. Female and first-winter male have pale throat and submoustachial stripe, dark malar stripe, and often have spotting on breast. Song consists of variations on repeated *chillia-chillia* warbles. Frequent summer visitor, breeding in more open lower Himalayan valleys, preferring terraced cultivation, orchards or forest clearings, 1200–2400 m. Winters in adjacent plains and western borders from Baluchistan to NWFP.

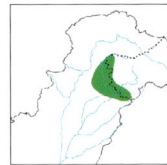

4 **GREY-WINGED BLACKBIRD** *Turdus boulboul*   28 cm

a MALE and **b** FEMALE Male black, with greyish wing-panel. Female olive-brown, with pale rufous-brown wing-panel. The finest songster among the sub-continent's thrushes. Locally common resident, in Himalayan temperate forest with good admixture of deciduous trees, 1500–2100 m. Confined to Murree Hills and a small disjunct population in the lower Kaghan Valley. Limited altitudinal migration in winter to better wooded, shady places.

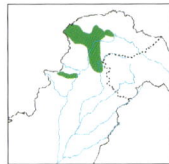

5 **EURASIAN BLACKBIRD** *Turdus merula*   25–28 cm

a MALE and **b** FEMALE Male is entirely black and female mainly brown, lacking pale wing-panel of Grey-winged. Flight call is a grating rattle. Song is a few spaced, mournful whistles. Uncommon resident, confined to alpine slopes above the tree-line, 3200–3900 m. Frequents grassy slopes and tumbled screes with patches of dwarf juniper, from Chitral eastwards to the Neelum (Kishenjanga) Valley. Descends to the tree-line in winter.

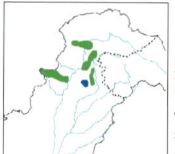

1 **CHESTNUT THRUSH** *Turdus rubrocanus* 27 cm
a MALE and **b** FEMALE Greyish head with chestnut upperparts and underparts. Male has pale collar; female has more uniform brownish-grey head/neck. Lengthy singing bouts consist of the repeated phrases *yee-bre-yee-bre-yee-bre-did-diyit-diddiyit-diddiyit-yip-bru-yipbru*. Locally common resident between 2300 m and 3300 m, wherever Himalayan forest with good understorey occurs; avoids the drier sunnier slopes that are so favoured by the Blue-capped Rock Thrush. Limited altitudinal movement in winter, a few individuals occurring in shadier ravines of the foothills.

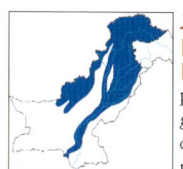

2 **DARK-THROATED THRUSH** *Turdus ruficollis* 25 cm
a MALE, **b** FEMALE and **c** 1ST-WINTER FEMALE *T. r. ruficollis*, and **d** MALE, **e** FEMALE and **f** 1ST-WINTER FEMALE *T. r. atrogularis* Uniform grey upperparts and wings. *T. r. atrogularis* has black throat and/or breast; first-winter has grey streaking on breast and flanks. *T. r. ruficollis*, which has been recorded on a couple of occasions, has red throat and/or breast, and red on tail; first-winter has rufous wash to supercilium and breast. Contact calls are a squeaky *qui-kweea*. Abundant winter visitor, arriving in northern mountainous regions mid-October and spreading slowly across NWFP, Baluchistan and Indus plains; mostly in Punjab by mid-November. A few linger in northern forested hills until mid-May.

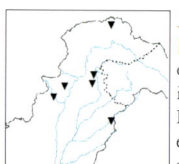

3 **DUSKY THRUSH** *Turdus naumanni* 24 cm
a MALE and **b** 1ST-WINTER Prominent supercilium contrasting with dark crown and ear-coverts; spotting across breast, forming breast-band and continuing down flanks (and contrasting with white belly), and largely chestnut wings. First-winter birds can be very dull, with chestnut in wings sometimes not apparent; in this plumage, broader supercilium and spotting on flanks are best features separating it from Dark-throated. Rare winter straggler to northern mountain valleys and adjacent plains, often in company with Dark-throated Thrushes.

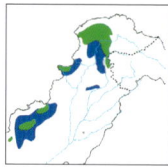

4 **MISTLE THRUSH** *Turdus viscivorus* 27 cm
ADULT Large size, pale grey-brown upperparts, whitish edges to wing feathers, and spotted breast. (See Appendix for comparison with Song Thrush.) Rather monotonous song comprises repeated phrases. Widespread resident in drier mountain regions wherever scattered forest occurs, from northern Baluchistan, Safed Koh and Chitral eastwards. Absent from Murree Hills except on passage. Nests at 1800–3600 m in orchards, tall coniferous forest and stunted juniper.

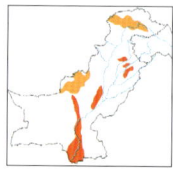

1 **SPOTTED FLYCATCHER** *Muscicapa striata*   15 cm
ADULT Large size, large and mainly dark bill, streaking on forehead and crown, indistinct eye-ring, and streaked throat and breast. A relatively silent bird, the breeding male occasionally giving spaced, brief, squeaky notes. Summer passage migrant from East Africa, with part of population staying to breed in drier mountainous regions of Baluchistan and main valleys in northern mountain areas, from Chitral to Baltistan, where locally common.

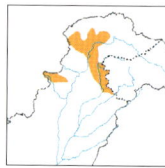

2 **DARK-SIDED FLYCATCHER** *Muscicapa sibirica*   14 cm
ADULT Small dark bill, and long primary projection. Breast and flanks heavily marked, with narrow pale line down centre of belly. Brief song is a complicated series of high-pitched repetitive trills and whistles. A common and widespread breeding bird in higher Himalayan forest, 2600–3350 m, both in Safed Koh and from Chitral to the Murree Hills. Surprisingly few winter records, but appears to remain in outer foothills. Hunts and nests around conifer tree canopy.

3 **ASIAN BROWN FLYCATCHER** *Muscicapa dauurica*   13 cm
ADULT Large bill with prominent pale base to lower mandible. Shorter primary projection than Dark-sided, and lores are more extensively pale than in that species. Pale underparts, with light brownish wash to breast and flanks. Song is a rather weak series of repeated trills and whistling phrases. Rare and extremely local summer breeding visitor to subtropical Pine zone, 1000–1300 m, southeast of the Murree Hills range.

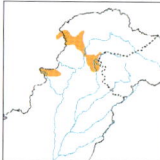

4 **RUSTY-TAILED FLYCATCHER** *Muscicapa ruficauda*   14 cm
ADULT Rufous uppertail-coverts and tail, rather plain face, and pale orange lower mandible. Brief song is louder and more melodious than other *Muscicapa* flycatchers, lasting about 3 seconds, *tree-e-loo-prili-prili-prit-prit-teeyou*. Passage migrant to central Asia, with locally frequent summer visitors breeding in Safed Koh, lower Chitral, and east through Dir, Swat and across to the Neelum Valley. Prefers more open forest and lower elevations than Dark-sided, 1500–3000 m.

5 **RED-THROATED FLYCATCHER**
    Ficedula parva   11.5–12.5 cm
a MALE and **b** FEMALE White sides to tail. Male has reddish-orange throat. Female and many males have creamy-white underparts. Very common double passage migrant throughout Pakistan, except in the eastern desert borders and southwest Baluchistan. A few overwinter in shady gardens, irrigated forest plantations and patches of riverine forest; the majority go further southeast into India.

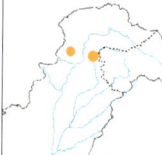

6 **KASHMIR FLYCATCHER** *Ficedula subrubra*   13 cm
a MALE and **b** FEMALE Male has more extensive and deeper red on underparts than Red-throated, with diffuse black border to throat and breast. Female and first-winter male can resemble some male Red-throated, but coloration of throat is more rufous, and this coloration is often more pronounced on breast than throat and often continues as wash onto belly and/or flanks. Rare and local spring passage migrant from Sri Lanka, with a few pairs breeding in watershed of the Neelum (Kishenjanga) Valley, and one record from lower Chitral. Prefers deciduous forest, 1800–2300 m. Globally threatened (Vulnerable).

FLYCATCHERS, PLATE 62

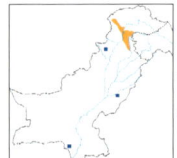

1 ULTRAMARINE FLYCATCHER *Ficedula superciliaris*   12 cm
a MALE, **b** FEMALE and **c** 1ST-SUMMER MALE   Male has deep blue upperparts and sides of neck/breast, and white underparts. Female has greyish-brown breast-side patches and lacks rufous on rump/uppertail-coverts. Territorial calls are a constantly repeated *chee-tr-r-r-r*; song consists of weak, high-pitched, disjointed trills and chirps. Locally common summer migrant, breeding in Himalayan moist temperate forest, 1800–2700 m. Most numerous in Murree Hills and lower Kaghan Valley, but extending westwards into Swat and lower Chitral. Winters mostly in peninsular India, but there are a few records of birds overwintering in shady gardens in Bahawalpur and Karachi.

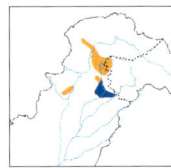

2 SLATY-BLUE FLYCATCHER *Ficedula tricolor*   13 cm
a MALE and **b** FEMALE   Male has white throat and white on tail; belly and flanks greyish. Female has white throat and rufous tail. Alarm call is a rapid *tek-tek-tek*, and song is a series of 3 or 4 high-pitched whistles, the last 2 rather hoarse and trilling. Resident but less common then Ultramarine, and favouring the same habitat but invariably at higher altitudes, 2600–8500 m. Prefers steeper open slopes on edge of forest. Winters in the immediate foothill regions.

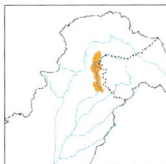

3 VERDITER FLYCATCHER *Eumyias thalassina*   15 cm
a MALE and **b** FEMALE   Male greenish blue, with black lores. Female duller and greyer, with dusky lores. Sings from top of tall trees, a rapid elided trilling, rising and falling in cadence. Locally common summer breeder, confined to Murree Hills, lower Kaghan Valley and lower parts of Neelum (Kishenjanga) Valley. Prefers tall trees on edge of forest glades, 1550–3200 m. Winters mostly in South India.

4 RUFOUS-BELLIED NILTAVA *Niltava sundara*   18 cm
a MALE and **b** FEMALE   Male has brilliant blue crown and neck-patch, and orange on underparts extending to vent. Female has oval-shaped throat-patch. A poor singer, starting with its warning call followed by 4 or 5 warbling trills, *sweee-tri-tri-tr-tih*. Scarce summer breeder, favouring damper, shadier slopes in Himalayan moist temperate forest, 1800–2600 m. Most frequent in Murree Hills and lower forested slopes of Kaghan Valley. Winters in foothills.

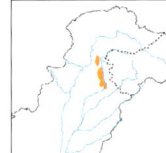

5 BLUE-THROATED FLYCATCHER
Cyornis rubeculoides   14 cm
a MALE and **b** FEMALE   Male has blue throat (some with orange wedge) and well-defined white belly and flanks. Female has creamy-orange throat, orange breast well demarcated from white belly, and creamy lores. Tails is rufescent brown. A low-altitude summer breeder, 460–900 m. Favours ravines with thickest cover, from the Margalla Hills eastwards through Lehtrar, Kahuta and Poonch. Like the Rufous-bellied Niltava, it moves eastwards in winter into the Siwaliks in India. Variable song comprises mimicry, rapid *Hippolais*-like churring, interspersed with very characteristic 3-noted rising and falling strophes, *turr-treee-turh*.

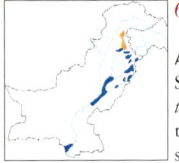

6 GREY-HEADED CANARY FLYCATCHER
Culicicapa ceylonensis   13 cm
ADULT   Grey head and breast, rest of underparts yellow, and greenish upperparts. Song comprises quite melodious whistling notes in rising and falling couplets, *twoi-toi-teeh-deeh*. Frequent summer breeder west of the Indus, in Himalayan moist temperate forest at lower altitudes, 1100–2100 m. Favours shady ravines with small streams. Winters over most of northern India, with numbers staying in Pakistan, frequenting orchards and old bungalow gardens in the plains.

ROBINS, PLATE 63

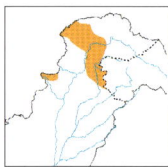

1 WHITE-TAILED RUBYTHROAT *Luscinia pectoralis*   14 cm
a MALE and **b** FEMALE Male has ruby-red throat, black breast-band and white on tail. Female has grey upperparts, grey breast-band and white tip to tail. Breeding males sing from dawn to dusk, often a prolonged and complicated series of rising and falling twitters, trills and warbles. Locally common resident, breeding in moister alpine regions, from Chitral eastwards to Baltistan, 2700–4500 m. Favours tumbled rocky screes with dwarf juniper. Migrates to adjacent foothills in winter. Absent from Murree Hills, rare in Safed Koh.

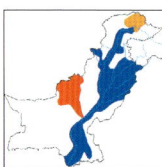

2 BLUETHROAT *Luscinia svecica*   15 cm
a MALE NON-BREEDING and **b** 1ST-WINTER FEMALE *L. s. svecica*, and **c** MALE BREEDING *L. s. abbotti* White supercilium and rufous tail sides. Male has variable blue, black and rufous patterning to throat and breast (patterning obscured by whitish fringes in fresh plumage). Female is less brightly coloured but usually with blue and rufous breast-bands. First-winter female may have just black submoustachial stripe and band of black spotting across breast. Breeding males have short, loud song with metallic *ting-ting-ting* and rich *torr-torr-torr* phrases. Common wintering bird throughout Indus plains, favouring proximity to water, whether canal-side seepage, lake margins or even sugarcane crops. A small population breeds in eastern Gilgit and Baltistan, particularly on the Deosai Plateau, favouring good scrub cover alongside streams and run-off gullies.

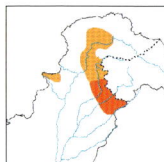

3 INDIAN BLUE ROBIN *Luscinia brunnea*   15 cm
a MALE and **b** FEMALE Male has blue upperparts and orange underparts, with white supercilium and black ear-coverts. Female has olive-brown upperparts, and buffish underparts with white throat and belly. Song, uttered continuously, comprises 3 or 4 piercing whistles decelerando, followed by rapid, tumbling *tit-tit-thwit-tichu-chuchu-cheeh*. Common summer breeder in Himalayan temperate forest with good understorey of shrubs, 2600–3200 m. Favours forest glades with rank vegetation and rarely shows itself. Winters in peninsular India and Sri Lanka in forest, scrub, and tea and coffee plantations.

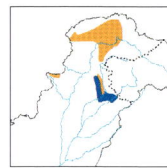

4 ORANGE-FLANKED BUSH ROBIN *Tarsiger cyanurus*   14 cm
a MALE and **b** FEMALE White throat, orange flanks, blue tail and redstart-like stance. Male has blue upperparts and breast sides. Female has olive-brown upperparts and breast sides. Breeding pairs repeat frog-like alarm croaks, and male's song is a plaintive, reedy *churrrh-chee*, monotonously repeated. Common resident, breeding near the upper limit of the tree-line and penetrating to the inner drier Himalayan forest areas, 2400–3350 m. Winters in adjacent foothills and lower valleys.

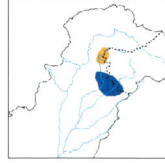

5 GOLDEN BUSH ROBIN *Tarsiger chrysaeus*   15 cm
a MALE and **b** FEMALE Orange to orange-buff underparts, with orange tail-sides. Pale legs. Male has broad orange supercilium, dark mask and orange scapulars. Female duller, with less distinct supercilium. A very rare local resident, recorded only recently from higher slopes on the edge of the tree-line in the Kaghan Valley, 2700–3350 m. Winters in foothills and recorded in October on the summit ridge (900 m) of the Margalla Hills.

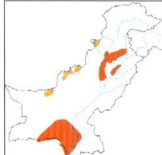

6 RUFOUS-TAILED SCRUB ROBIN
Cercotrichas galactotes   15 cm
ADULT Long rufous tail, tipped white and with black subterminal markings. Sandy-grey upperparts and creamy-white underparts, with whitish supercilium and dark eye-stripe and moustachial stripe. Nesting males sing in fluttering flight display and from bush tops, sweet melodious notes in a rising and falling scale. Scarce summer breeder in bush-dotted run-off channels and non-perennial stream beds in Baluchistan and NWFP. Winters in East Africa, with notable autumn passage through Sind on the west bank of the Indus.

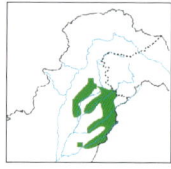

1 **ORIENTAL MAGPIE ROBIN** *Copsychus saularis* 23 cm
a MALE and **b** FEMALE Black/slate-grey and white, with white on wing and at sides of tail. Scarce and local resident on western extremity of range. Mainly northern Punjab, in shadier gardens, irrigated plantations and eroded foothills up to 1500 m.

2 **INDIAN ROBIN** *Saxicoloides fulicata* 19 cm
a MALE and **b** FEMALE Reddish vent and black tail in all plumages. Male has white shoulders and black underparts. Female has greyish underparts. Short plaintive song comprises high-pitched whistling warble. Common widespread resident, avoiding forested areas and urban settlements, and preferring arid stony areas.

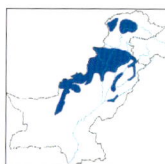

3 **RUFOUS-BACKED REDSTART** *Phoenicurus erythronota* 16 cm
a MALE BREEDING, **b** 1ST-WINTER MALE and **c** FEMALE Large size. Often holds tail slightly cocked. Male has rufous mantle and throat, white on wing, and black mask; plumage heavily obscured by pale fringes in non-breeding and first-winter plumages. Female has double buffish wing-bar. Locally frequent winter visitor, mainly to mountainous regions from central Baluchistan (up to 2400 m), through Chitral and Gilgit, often with snow on the ground. Also northern Punjab, especially Salt Range.

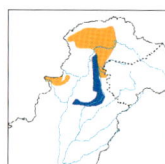

4 **BLUE-CAPPED REDSTART**
Phoenicurus coeruleocephalus 15 cm
a MALE BREEDING, **b** FEMALE and **c** 1ST-WINTER MALE Male has blue-grey cap, black tail, and white on wing; the dark areas have pale fringes in non-breeding and first-winter plumages. Female has grey underparts, prominent double wing-bar, blackish tail and chestnut rump. Alarm call is a rapid *tik-tik*. Song comprises pleasant warbling see-saw phrases *trrri-trrru-trrri-trrru*. Locally common resident, summering in inner higher Himalayan forest areas, 2300–3700 m. Prefers forested valleys in far northern areas, and upper limit of tree-line in Hazara District. Absent from Murree Hills. Altitudinal migrant to foothills and, occasionally, Punjab Salt Range.

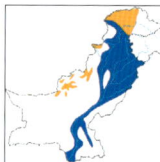

5 **BLACK REDSTART** *Phoenicurus ochruros* 15 cm
a MALE and **b** FEMALE Male has black throat and breast, with grey crown becoming almost white on forehead. Underparts rufous. Female and first-year male are mainly dusky brown with rufous tail. Short warbling song terminates in interrogative whistles and an explosive hissing rattle. Widespread common winter visitor throughout Indus plains, preferring tree-lined canals and old gardens. Summer breeder in juniper and Chilghoza Pine forest, Baluchistan and open rocky alpine areas of the far north, up to the lower Khunjerab Valley, 3000–4250 m.

6 **COMMON REDSTART** *Phoenicurus phoenicurus* 15 cm
a MALE BREEDING, **b** 1ST-WINTER MALE and **c** FEMALE Told from male Black Redstart by paler grey upperparts, black confined to throat, and paler, more orange coloration to underparts; in non-breeding and first-winter plumages, coloration is obscured by pale feather fringes (when appearance is quite different to any Black Redstart). Female has buff-brown upperparts and buffish underparts (and is paler and more warmly coloured than female Black). Restricted passage migrant along extreme western borders, southern Baluchistan and northern Chitral.

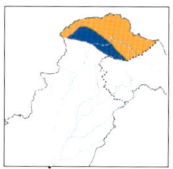

7 **WHITE-WINGED REDSTART**
Phoenicurus erythrogaster 18 cm
a MALE and **b** FEMALE Large size and stocky appearance. Male has white cap and large white patch on wing. Female has buff-brown upperparts and buffish underparts. Hardy resident, breeding in far northern regions up to 4800 m. Winters down to 1750 m in main valleys in the far north. Restless summer insectivorous foraging; in winter, groups defend Sea Buckthorn patches.

1 **BLUE-FRONTED REDSTART** *Phoenicurus frontalis*   15 cm
a MALE and **b** FEMALE Orange rump and tail sides, with black centre and tip to tail. Male has blue head and upperparts, and chestnut-orange underparts; coloration heavily obscured by rufous-brown feather fringes in non-breeding and first-winter plumages. Female has dark brown upperparts and underparts, with orange wash to belly; tail pattern is the best feature separating it from other female redstarts. Alarm calls are a *ee-tit*, and the short song comprises 2 prolonged trills followed by rising and falling whistling phrases. Locally frequent resident, breeding at the upper limit of the tree-line in summer, down to adjacent valleys in winter. Prefers damper bush-lined ravines and Himalayan birch scrub for nesting. Does not descend as far as the foothills in winter (1000 m).

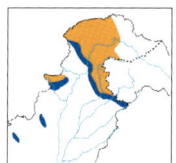

2 **WHITE-CAPPED WATER REDSTART**
Chaimarrornis leucocephalus   19 cm
ADULT White cap, and rufous tail with broad black terminal band. Mainly mountain streams and rivers. Contact call is a penetrating *tseet*. Brief song is a rather weak undulating, whistling *tieu-yieu-yieu-yieu*. Locally common resident from Chitral eastwards to Baltistan. Confined to mountain streams and rivers up to 3600 m, wintering down to adjacent plains and foothills, often by tiny streams, with some individuals staying in inner valleys in northern areas.

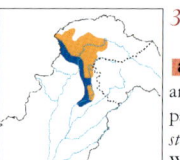

3 **PLUMBEOUS WATER REDSTART**
Rhyacornis fuliginosus   12 cm
a MALE and **b** FEMALE Male is slaty blue, with rufous-chestnut tail. Female and first-year male have black-and-white tail and white spotting on grey underparts. Alarm call is a strident *peet-peet*, and song is a short, rapid, stridulating *streee-treee-treee*. Mountain streams and rivers. Common resident, sympatric with White-capped Water Redstart, but adapted to smaller streams and is a less adventuresome altitudinal winter migrant, mostly in streams down to 1800 m. Absent from the Kurram Valley but common from Chitral down to the Murree Hills.

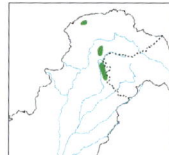

4 **WHITE-BELLIED REDSTART**
Hodgsonius phaenicuroides   18 cm
a MALE and **b** FEMALE Long, graduated tail that is often held cocked and fanned. Male has white belly, rufous tail sides, and white spots on alula. Female has white throat and belly, and chestnut on tail. Warning call is a rapid grating croak. Brief song comprises 3 or 4 wistful whistling phrases, *teuuh-tiyou-tiyou-tuh*. Breeds in subalpine shrubberies; winters in thick undergrowth and forest edges at lower levels. Scarce resident, with small disjunct population, mainly in Hazara District and the Neelum Valley (2400–3400 m). Prefers open slopes with thick cover of *Viburnum*, *Sambucus* and *Paeonia*. Limited winter altitudinal migration.

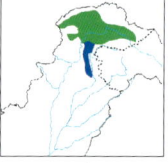

5 **LITTLE FORKTAIL** *Enicurus scouleri*   12 cm
ADULT Small and plump, with a short tail. Black upperparts with white forehead and wing-bar. A relatively silent species. Nest cleverly concealed on wet mossy cliffs and made of moss. Mountain streams; also slower-moving streams in winter. Locally frequent resident over a wide range of mountain torrents and major rivers, penetrating drier, less forested valleys than Spotted Forktail, from Chitral to Baltistan and southwards to Murree Hills and Neelum Valley (3000 m). Winters mainly in broader mountain valleys, with a few down to the foothills.

6 **SPOTTED FORKTAIL** *Enicurus maculatus*   25 cm
ADULT Large size and very long tail; white forehead, white spotting on mantle, and black breast. Alarm call is a loud *cheet-cheet*. Song comprises trills and bubbling warbles interspersed with harsher *dzeet-dzit* calls. Frequent resident, confined to streams and rivers in moist forested areas, avoiding open country, from southern Chitral southwards to Murree Hills and Neelum Valley. Not recorded in the Kurram Valley or beyond southwestern Gilgit. A few straggle down to foothill streams in winter.

CHATS AND WHEATEARS, PLATE 66

1 **STOLICZKA'S BUSHCHAT** *Saxicola macrorhyncha*   17 cm
a MALE BREEDING, **b** MALE NON-BREEDING and **c** FEMALE Male has white supercilium, white primary coverts and much white on tail; upperparts and ear-coverts appear blackish when worn (breeding) and streaked when fresh (non-breeding). Female differs from female Common Stonechat in longer bill and tail, more prominent supercilium, and broad buffish edges and tips to tail feathers. Believed extinct in Pakistan. No records since 1922. A sedentary species confined to scrub desert. Globally threatened (Vulnerable).

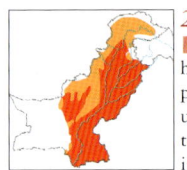

2 **COMMON STONECHAT** *Saxicola torquata*   17 cm
a MALE BREEDING, **b** MALE NON-BREEDING and **c** FEMALE Male has black head, white patch on neck, orange breast and whitish rump (features obscured by pale feather fringes in fresh plumage); lacks white in tail. Female has streaked upperparts and orange on breast and rump. Tail darker than in female White-tailed. Widespread and common resident, breeding over wide altitudinal range in northern Baluchistan, NWFP and throughout northern areas (1500–3880 m). Prefers open unforested slopes or wider valleys for breeding, wintering mostly in peninsular India, with strong April passage. A few overwinter in uncultivated tracts in Sind and Punjab.

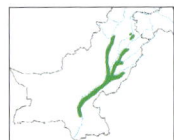

3 **WHITE-TAILED STONECHAT** *Saxicola leucura*   12.5–13 cm
a MALE BREEDING and **b** FEMALE Similar in plumage to Common Stonechat. Male has largely white inner webs to tail feathers. Female has greyer upperparts than Common, with diffuse streaking, and paler grey-brown tail. Rare and sedentary resident confined to extensive *Saccharum* (cane grass) tracts along major rivers, from Sind to Punjab, and Khattak District in NWFP.

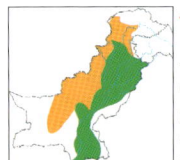

4 **PIED BUSHCHAT** *Saxicola caprata*   13.5 cm
a MALE BREEDING, **b** FEMALE and **c** 1ST-WINTER MALE Male black, with white rump and wing-patch; rufous fringes to body in non-breeding and first-winter plumages. Female has dark brown upperparts and rufous-brown underparts, with rufous-orange rump. Short song starts with piercing whistles, then rising and falling warbles. Very common resident in Indus basin, in open treeless tracts with patches of cane grass. Avoids forest and environs of human habitation. Breeds in Indus plains and is a summer migrant into broader northern mountain valleys (300–2400 m).

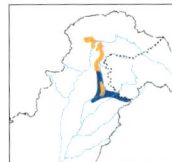

5 **GREY BUSHCHAT** *Saxicola ferrea*   15 cm
a MALE and **b** FEMALE Male has white supercilium and dark mask; upperparts grey to almost black, depending on extent of feather wear. Female has buff supercilium and rufous rump and tail sides. Brief song, *titteratu-chak-tew-titatit*. Warning call is a soft *zizzing*. Common resident in moister forested areas of outer Himalayas, with limited altitudinal winter migration. Breeds in open glades and forest edges (1900–3000 m).

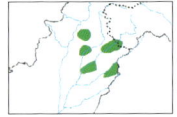

6 **BROWN ROCK-CHAT** *Cercomela fusca*   17 cm
ADULT Both sexes brown, with more rufescent underparts. Scarce and locally distributed resident confined to rocky outcrops or large-scale masonry buildings in northern Punjab.

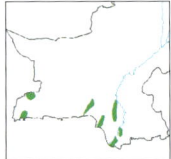

7 **HOODED WHEATEAR** *Oenanthe monacha*   17.5 cm
a ADULT, **b** 1ST-WINTER MALE and **c** FEMALE Long bill. Male has white crown and largely white outer tail feathers; in winter, buffish or greyish wash to crown and pale fringes to upperparts and wings (first-winter male has tail as female). Female has rufous-buff rump and tail, with brown central tail feathers. Rare and local resident to the west bank of the Indus along the Mekran coast, preferring the most barren tracts.

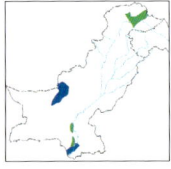

8 **HUME'S WHEATEAR** *Oenanthe alboniger*   17 cm
ADULT All-black head and largely white underparts. Told from *picata* race of Variable by stockier appearance and domed head, larger bill, glossy sheen to black of plumage (except when worn), black of throat not extending so far down on breast, and white of rump extending farther up lower back. Frequent sedentary resident on the west bank of the Indus, lower Sind, Las Bela in rocky desert, and the same habitat in eastern Gilgit and Baltistan. Both sexes are alike.

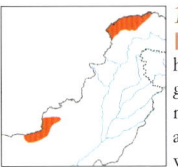

1 **NORTHERN WHEATEAR** *Oenanthe oenanthe*   15 cm
a MALE BREEDING, **b** FEMALE BREEDING and **c** 1ST-WINTER Breeding male has blue-grey upperparts, black mask and pale orange breast. Breeding female is greyish to olive-brown above; lacks rufous patch on ear-coverts of Finsch's and never shows dark grey/black on throat. Compared with Isabelline, adult winter and first-winter have blackish centres to wing-coverts and tertials, and show more white at sides of tail. Rare spring passage migrant along the extreme western borders of Baluchistan and Chitral. Not recorded on return passage.

2 **FINSCH'S WHEATEAR** *Oenanthe finschii*   14 cm
a MALE, **b** FEMALE and **c** 1ST-WINTER FEMALE Male has creamy-buff to white mantle. Adult female and first-winter have rufous cast to ear-coverts, rather pale grey-brown upperparts, and broad black terminal band to tail. Frequent winter visitor, October to March, west-central Baluchistan, favouring desolate boulder-strewn slopes. Occasionally breeds at Chagai near the Afghan border.

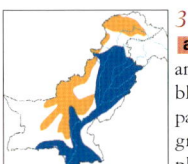

3 **VARIABLE WHEATEAR** *Oenanthe picata*   14.5 cm
a MALE and **b** FEMALE *O. p. picata*, **c** MALE and **d** FEMALE *O. p. capistrata*, and **e** MALE and **f** FEMALE *O. p. opistholeuca*. Very variable. Males can be mainly black, have a black head with white underparts, or a white crown and white underparts. Females can be mainly sooty brown or have greyish upperparts with variable greyish-white underparts. Commonest resident wheatear, wintering in Indus plains and foothill country. Scattered breeding throughout arid stony slopes, and valleys from Baluchistan north to Chitral.

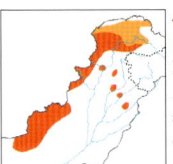

4 **PIED WHEATEAR** *Oenanthe pleschanka*   14.5 cm
a MALE, **b** FEMALE, **c** 1ST-WINTER MALE and **d** 1ST-WINTER FEMALE Different tail pattern from Variable: always shows black edge to outer feathers (lacking in Variable) and often has only a narrow and broken terminal black band (broad and even on Variable). On breeding male, white of nape extends to mantle, black of throat does not extend to upper breast, and breast is washed with buff (features separating it from *capistrata* race of Variable). Non-breeding and first-winter male and female have pale fringes to upperparts and wings (not apparent on fresh-plumaged Variable). Frequent double passage migrant along the western borders of Baluchistan and NWFP. Summer breeding in Chitral, east to Baltistan (2700–4100 m). Favours broad valleys and open stony slopes.

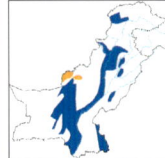

5 **RUFOUS-TAILED WHEATEAR**
Oenanthe xanthoprymna   14.5 cm
a MALE BREEDING and **b** MALE NON-BREEDING Rufous-orange lower back and rump, and rufous tail sides. Male has black lores. Summers on dry rocky slopes; winters in semi-desert. Uncommon winter visitor, favouring rocky gorges, ravines west of the Indus, bare salt pans, and barren scrub desert in Sind and Punjab. Breeds sparingly in northwestern Baluchistan.

6 **DESERT WHEATEAR** *Oenanthe deserti*   14–15 cm
a MALE BREEDING, **b** MALE NON-BREEDING, **c** FEMALE BREEDING and **d** FEMALE NON-BREEDING Sandy-brown upperparts, with largely black tail and contrasting white rump. Male has black throat (partly obscured by pale fringes in fresh plumage). Female has blackish centres to wing-coverts and tertials in fresh plumage, and largely black wings when worn (useful distinction from Isabelline). Widespread common winter visitor throughout more barren tracts at lower altitudes in Baluchistan, and avoiding cultivated areas in Sind and Punjab. Breeding records in the Quetta Valley at the beginning of the 20th century.

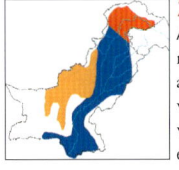

7 **ISABELLINE WHEATEAR** *Oenanthe isabellina*   16.5 cm
ADULT Rather plain sandy-brown and buff. Tail shorter than in Desert, and with more white at base and sides. Has paler, sandy-brown wings with contrasting dark alula (lacking black centres to coverts and tertials/secondaries). Singing male is a wonderful mimic. Common winter visitor throughout the Indus basin, breeding widely in Baluchistan and NWFP, favouring wide stony plateaux or cultivation edges (900–1500 m).

STARLINGS AND MYNAS, PLATE 68

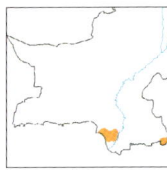

1 **CHESTNUT-TAILED STARLING** *Sturnus malabaricus*   20 cm
a ADULT and **b** JUVENILE Adult has grey upperparts, rufous underparts and chestnut tail. Juvenile is rather uniform, with rufous sides and tips to outer tail feathers. Uncommon, with irregular irruptions to extreme southeastern Sind, mainly in early spring.

2 **BRAHMINY STARLING** *Sturnus pagodarum*   21 cm
a ADULT and **b** JUVENILE Adult has black crest, and rufous-orange sides of head and underparts. Juvenile lacks crest; has grey-brown cap and paler orange-buff underparts. Uncommon resident in northern foothill areas, with some migrating to main valleys of Chitral, Swat and Gilgit for breeding (900–1800 m). There is a small resident population near Karachi. Favours terraced cultivation with orchards.

3 **ROSY STARLING** *Sturnus roseus*   21 cm
a ADULT and **b** JUVENILE Adult has blackish head with shaggy crest, pinkish mantle and underparts, and blue-green gloss to wings. In non-breeding and first-winter plumages it is much duller, the pink of the plumage partly obscured by buff fringes and the black by greyish fringes. Juvenile is mainly sandy brown, with stout yellowish bill, and broad pale fringes to wing feathers. Abundant double passage migrant, mainly through NWFP and Punjab, and lower Baluchistan and Sind. Highly gregarious. First autumn arrivals appear in Sind in late July.

4 **COMMON STARLING** *Sturnus vulgaris*   21 cm
a ADULT BREEDING, **b** ADULT NON-BREEDING and **c** JUVENILE Adult is metallic green and purple; heavily marked with buff and white in winter. Juvenile is dusky brown with whiter throat. Abundant gregarious winter visitor across much of Pakistan, favouring grassy lake margins and cultivation, with small sedentary breeding population (depicted in map) associated with major irrigation canals and river-barrage seepage zones.

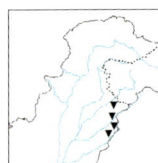

5 **ASIAN PIED STARLING** *Sturnus contra*   23 cm
a ADULT and **b** JUVENILE Adult is black and white, with orange orbital skin and a large, pointed, yellowish bill. Juvenile has brown plumage in place of black. Uncommon recent colonisers in northwest Punjab, favouring damper, well-wooded areas.

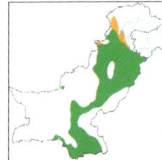

6 **COMMON MYNA** *Acridotheres tristis*   25 cm
a b ADULT Brownish myna with yellow orbital skin, white wing-patch and white tail-tip. Juvenile duller. Very abundant semi-commensal resident, avoiding only remoter mountain areas or extensive desert tracts. Increased urbanisation has favoured this species.

7 **BANK MYNA** *Acridotheres ginginianus*   23 cm
a b ADULT Orange-red orbital patch, orange-yellow bill, and tufted forehead. Wing-patch, underwing-coverts and tail-tip orange-buff. Adult is bluish grey with blackish cap. Juvenile is duller and browner than adult. Colonial nesting resident in Sind, Punjab and cultivated tracts in NWFP. Avoids desert or hilly tracts and most abundant rice-growing areas. Often associated with grazing animals.

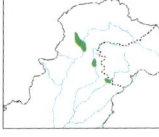

8 **JUNGLE MYNA** *Acridotheres fuscus*   23 cm
a ADULT and **b** JUVENILE Tufted forehead, and white wing-patch and tail-tip; lacks bare orbital skin. Juvenile is browner, with reduced forehead tuft. Rather rare and local resident, confined to cultivated mountain valleys in Indus Kohistan, Hazara tribal territory and Swat. Suffers competition from the more aggressive Common Myna in eastern former parts of its range.

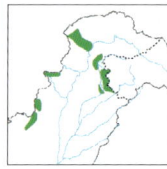

1 **KASHMIR NUTHATCH** *Sitta cashmirensis*   12 cm
a MALE and **b** FEMALE Compared with female Chestnut-bellied, has uniform undertail-coverts and lacks clearly defined white cheeks. Distinctive rasping jay-like call. Locally frequent resident, with scattered disjunct populations from Chilghoza Pine forest in northern Baluchistan to Chitral, Swat and down to the Murree Hills (1800–3350 m). Prefers lower altitudes and predominantly deciduous forest in contrast to the White-cheeked Nuthatch.

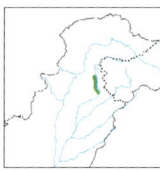

2 **CHESTNUT-BELLIED NUTHATCH** *Sitta castanea*   12 cm
a MALE and **b** FEMALE Male has deep chestnut underparts and white cheeks. Female has paler cinnamon-brown on underparts, although its white cheeks are more pronounced than on similar species; has pale fringes to undertail-coverts. Calls include an explosive *siditit*. Rare and locally restricted resident in subtropical Himalayan foothill forest at the eastern end of the Murree Hills. Favours predominantly deciduous forest with *Salmalia*, *Pistacia* and *Mallotus* trees.

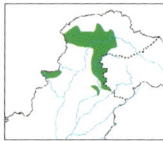

3 **WHITE-CHEEKED NUTHATCH** *Sitta leucopsis*   12 cm
ADULT Black crown and nape, white face, and whitish underparts with rufous flanks and undertail-coverts. Call is likened to the bleating of a young goat. A hardy common resident, from moist Himalayan temperate forest to inner dry coniferous forest, and with very little altitudinal movement in winter; such forests ring with their peculiar bleating contact calls (2300–3600 m).

4 **EASTERN ROCK NUTHATCH** *Sitta tephronota*   15 cm
ADULT Large, with very large bill. Long black eye-stripe, pale blue-grey upperparts and whitish underparts. Loud contact calls are a *twoi-twoi*. Common sedentary resident in western parts of Baluchistan, from barren plateaux valleys to juniper-dotted mountain slopes. Not recorded north of the Takht-e-Suleiman range.

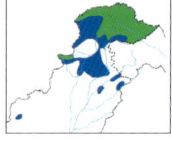

5 **WALLCREEPER** *Tichodroma muraria*   16 cm
a MALE BREEDING and **b** ADULT NON-BREEDING Long, down-curved bill. Largely crimson wing-coverts; shows white primary spots in flight. Breeding male has black throat. Frequent resident throughout far northern mountains, from Chitral across the Karakoram range, with considerable altitudinal migration to lower valleys and adjacent plains in winter.

6 **EURASIAN TREECREEPER** *Certhia familiaris*   12 cm
ADULT Unbarred tail, and whitish throat and breast. Mainly conifer/birch forest. Rare and patchy in distribution: inner drier Himalayan coniferous forest, Gilgit and northern Hazara District (2700–3350 m).

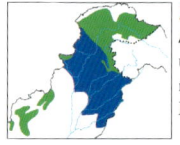

7 **BAR-TAILED TREECREEPER** *Certhia himalayana*   12 cm
ADULT Dark barring on tail, white throat, and dull whitish or dirty greyish-buff underparts. Common and widespread resident, from higher forested ranges in northern Baluchistan, Safed Koh and across northern areas, extending down to the Murree Hills. Winters on adjacent plains.

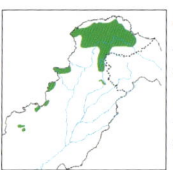

8 **WINTER WREN** *Troglodytes troglodytes*   9.5 cm
ADULT Small and squat, with stubby tail. Brown, with dark-barred wings, tail and underparts. Locally common resident, breeding up to 3800 m throughout northern areas and in isolated populations in higher Baluchistan and NWFP ranges. Winters down to 1000 m in main valleys. Favours rocky screes and the subalpine zone for breeding.

TITS, PLATE 70

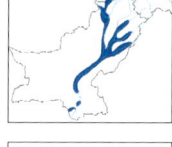

1 **WHITE-CROWNED PENDULINE TIT** *Remix coronatus*   10 cm
a FRESH MALE, **b** FEMALE and **c** JUVENILE Male has blackish mask and nape, whitish crown and whitish collar. Female has pale grey crown and collar, and lacks black nape-band. Juvenile lacks dark mask. A scarce winter visitor, entering via the Kurram Valley and favouring riverine forest in proximity to *Typha* reed-beds or irrigated forest plantations. Frequent in Punjab, absent from lower Sind.

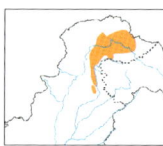

2 **FIRE-CAPPED TIT** *Cephalopyrus flammiceps*   10 cm
a MALE BREEDING, **b** MALE NON-BREEDING and **c** FEMALE Flowerpecker-like, with greenish upperparts and yellowish to whitish underparts. Lacks crest. Breeding male has bright orange-scarlet forecrown. Scarce summer breeder in the Murree Hills and Hazara District. Winters in peninsular India. Prefers Himalayan forest with a scattering of deciduous trees (1800–3000 m).

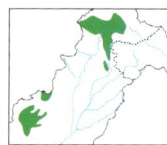

3 **RUFOUS-NAPED TIT** *Parus rufonuchalis*   13 cm
ADULT Large size, with extensive black bib (to upper belly) and grey belly. Lacks wing-bars. Common resident in central Baluchistan in juniper forest, and northern Himalayan forest from Chitral across Baltistan down to the Murree Hills. Tolerant of very cold conditions (1500–3200 m).

4 **SPOT-WINGED TIT** *Parus melanolophus*   11 cm
ADULT Small size, with broad white tips to median and greater coverts, blue-grey mantle, rufous breast sides and dark grey belly. Very common resident at 1800–3200 m in Himalayan coniferous forest, but avoids more xeric open habitat tolerated by the Rufous-naped Tit. Limited altitudinal winter movement to foothills.

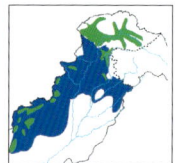

5 **GREAT TIT** *Parus major*   14 cm
ADULT Black breast centre and line down belly, greyish mantle, greyish-white breast sides and flanks, and white wing-bar. Juvenile has yellowish-white cheeks and underparts, and yellowish-olive wash to mantle. Widespread common resident, from higher ranges of Baluchistan, NWFP and main valleys in northern areas (1000–2400 m), and with widespread winter movement to adjacent foothills and plains across northern Punjab. Prefers open country at lower elevations.

6 **GREEN-BACKED TIT** *Parus monticolus*   12.5 cm
ADULT Green mantle and back, and yellow on underparts. Forest; prefers moister habitat than Great Tit. Common resident in moist temperate Himalayan forest with a good admixture of deciduous trees (1500–2700 m). A few straggle down to the Margallas and Punjab Salt Range in winter.

7 **BLACK-LORED TIT** *Parus xanthogenys*   13 cm
ADULT Prominent black crest, and black centre to yellow throat and breast. Black forehead and lores, greenish upperparts (with black streaking to scapulars) and yellowish wing-bars. Rare straggler to the outer Murree foothills and Margallas. Small resident population in Murree Hills in early 20th century now extinct.

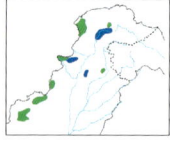

8 **WHITE-CHEEKED TIT** *Aegithalos leucogenys*   11 cm
ADULT Black throat, cinnamon crown, yellowish iris and grey-brown mantle. Juvenile has buffish-white throat and streaking on breast. Locally frequent resident in scattered populations in higher ranges of western Baluchistan, NWFP and Chitral to Baltistan (1500–3600 m). Prefers drier forested slopes with Holly Oak, juniper and Chilghoza Pine.

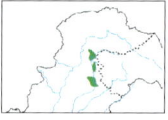

9 **BLACK-THROATED TIT** *Aegithalos concinnus*   10.5 cm
ADULT Chestnut crown, white chin and black throat, white cheeks and grey mantle. Juvenile has white throat and indistinct black-spotted breast-band. Locally frequent resident in moister lower Himalayan forest with predominantly deciduous trees, from Hazara District, the Neelum Valley and Murree Hills (1375–2400 m).

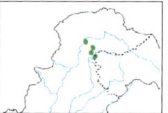

10 **WHITE-THROATED TIT** *Aegithalos niveogularis*   11 cm
ADULT White forehead and forecrown and whitish throat; iris brownish. Diffuse blackish mask and cinnamon underparts, with darker breast-band. Juvenile has dusky throat, more prominent breast-band, and paler lower breast and belly. Rare resident, confined to the upper limit of the tree-line and subalpine scrub, Hazara District, and Neelum Valley. Absent from northern areas and the Murree Hills.

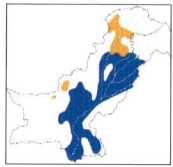

1 PALE MARTIN *Riparia diluta* — 13 cm
a b c ADULT Upperparts paler and greyer than on Plain Martin. Throat is greyish white and has a poorly defined breast-band. Locally migratory common resident, breeding colonially in major river valleys in northern Baluchistan, Chitral, Hazara District and south to the Punjab Salt Range (300–3000 m). Winters throughout the Indus basin, favouring lakes and rivers.

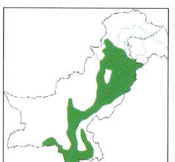

2 PLAIN MARTIN *Riparia paludicola* — 12 cm
a b c ADULT Upperparts and underwing darker than on Pale Martin. Pale brownish-grey throat and breast, merging into dingy-white rest of underparts; some have suggestion of breast-band. Locally abundant resident, confined to the Indus basin and avoiding mountainous regions, preferring the proximity of larger rivers and lakes.

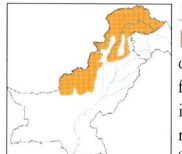

3 EURASIAN CRAG MARTIN *Hirundo rupestris* — 15 cm
a b c ADULT Larger and stockier than the *Riparia* martins. Dark underwing-coverts, dusky throat, and brown flanks and undertail-coverts (the latter with pale fringes). Lacks breast-band. Shows white spots in tail when spread. Frequent resident, favouring mountainous regions, especially in the vicinity of gorges. Locally migratory to breed non-colonially throughout northern areas (1300–4700 m). Sympatric with Pale Crag Martin in southern part of its range.

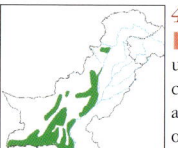

4 ROCK MARTIN *Hirundo fuligula* — 13 cm
a b c ADULT Smaller than Eurasian Crag Martin, with paler sandy-grey upperparts (especially rump), buffish-white throat, paler vent and undertail-coverts, and paler underwing-coverts. Uncommon sedentary resident, favouring arid rocky gorges and scrub desert, from the Mekran coast north to dry foothills on the west bank of the Indus and up to Dera Ghazi Khan.

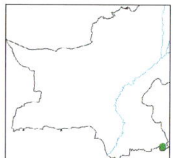

5 DUSKY CRAG MARTIN *Hirundo concolor* — 13 cm
a b c ADULT Upperparts and underparts dark brown and rather uniform; much darker in appearance than other crag martins. Rare and local resident, occurring only in the Kharunjhar Hills on the border of Rann of Kutch.

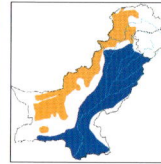

6 BARN SWALLOW *Hirundo rustica* — 18 cm
a b c ADULT and **d** JUVENILE Reddish throat, long tail-streamers and blue-black breast-band. Juvenile duller; lacks tail-streamers. Abundant resident, wintering throughout the Indus basin and summer breeding in foothills and at lower elevations throughout Baluchistan, north to southern Chitral. Often nests in occupied buildings (450–1800 m).

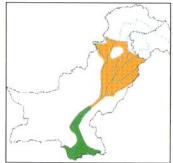

7 WIRE-TAILED SWALLOW *Hirundo smithii* — 14 cm
a b c ADULT and **d** JUVENILE Chestnut crown, white underparts and fine tail projections. Juvenile has brownish cast to blue upperparts, and dull brownish crown. Common summer migrant, breeding throughout the Indus basin up to 1800 m. Avoids mountainous areas, with the majority wintering in peninsular India, although some are resident in southern Sind.

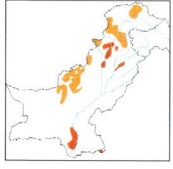

8 RED-RUMPED SWALLOW *Hirundo daurica* — 16–17 cm
a b c ADULT and **d** JUVENILE Rufous-orange neck sides and rump, finely streaked buffish-white underparts, and black undertail-coverts. Common and widespread summer visitor, breeding in mountainous tracts from Baluchistan to NWFP and throughout northern areas. Not so dependent on proximity to water as other swallows (450–3300 m).

1 **STREAK-THROATED SWALLOW** *Hirundo fluvicola*   11 cm
a b c ADULT and **d** JUVENILE Small, with slight fork to long, broad tail. Chestnut crown, streaked throat and breast, white mantle streaks and brownish rump. Juvenile is duller, with browner crown. Highly gregarious and sedentary resident across Indus plains, except southern Sind. Absent from mountainous areas. Spread of irrigation canals has favoured this species.

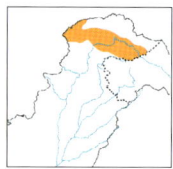

2 **NORTHERN HOUSE MARTIN** *Delichon urbica*   12 cm
a b c ADULT Told from Asian by whiter underparts and underwing-coverts, longer and more deeply forked tail, and whiter and more extensive rump patch. Locally common summer migrant, nesting in innermost northern valleys in Hunza, the Karakorams and Baltistan (3000–4500 m). Small numbers occur on passage south of the Himalayas.

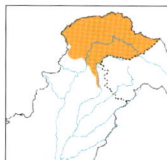

3 **ASIAN HOUSE MARTIN** *Delichon dasypus*   12 cm
a b c ADULT Dusky-white underparts and rump, shallow tail fork, dusky underwing and (not always) dusky centres to undertail-coverts. Common summer breeding migrant throughout northern areas, often nesting on alpine cliffs up to 3400 m, also in houses in hill villages. Winters mostly extra-limitally in India.

4 **WHITE-EARED BULBUL** *Pycnonotus leucotis*   20 cm
a ADULT *P. l. leucotis* and **b** ADULT *P. l. humii* White-cheeked bulbul with black crown and nape, and short (in northwest part of range) or non-existent crest. Very common widespread resident throughout Indus plains, extending westwards into cultivated regions of Baluchistan and NWFP, and eastwards throughout desert borders. Prefers more arid habitat than Red-vented Bulbul.

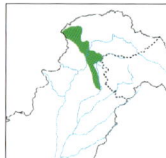

5 **HIMALAYAN BULBUL** *Pycnonotus leucogenys*   20 cm
ADULT Brown crest and nape, white cheeks with black throat, and yellow vent. Locally common resident in main valleys and foothills from Chitral and Dir to Swat, extending down to the Murree Hills and Margallas (1500–2100 m).

6 **RED-VENTED BULBUL** *Pycnonotus cafer*   20 cm
ADULT Blackish head with slight crest, scaled appearance to upperparts and breast, red vent and white rump. Abundant resident throughout the Indus basin, including urban areas; up to 1600 m in outer foothills, avoiding only desert tracts. Spread of irrigation has favoured this species at the expense of the White-eared Bulbul.

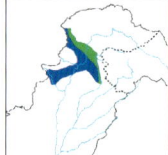

7 **BLACK BULBUL** *Hypsipetes leucocephalus*   25 cm
ADULT Slate-grey bulbul with black crest. Has shallow fork to tail, and red bill, legs and feet. Juvenile lacks crest; has whitish underparts with diffuse grey breast-band, and brownish cast to upperparts. Locally frequent resident in the outer foothills of the Himalayas where deciduous trees or orchards predominate (1000–2100 m), from lower Chitral and Hazara District to the Neelum Valley and Murree Hills.

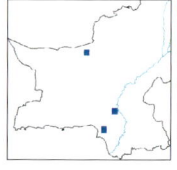

8 **GREY HYPOCOLIUS** *Hypocolius ampelinus*   25 cm
a MALE and **b** FEMALE Crested, with a long tail and white on the primaries. Male has a black mask and tail-band. Female rather uniform sandy brown. Highly nomadic and irregular rare visitor from southern Baluchistan to Sind Kohistan, always in scrub desert. Omnivorous, often foraging in flocks when food supplies are abundant after localised rain showers.

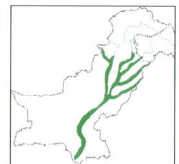

1 RUFOUS-VENTED PRINIA *Prinia burnesii*   17 cm

ADULT Large size, with streaked upperparts, whitish lores/eye-ring, broad tail and rufous vent. Found in quite different habitat to Striated; also separated from that species in that it lacks well-defined pale tips and dark subterminal bands to underside of tail (apparent in Striated), underparts are whiter, and it has rather uniform wings (lacking rufous edges to remiges in Striated). Warning calls are a *skeeow-skeeow*, and short song is a *tuweet-tiwittoo-tweeet-tuwittoo-weeto-witoo*. Resident, occurring frequently along margins of larger rivers/lakes and especially in irrigation-barrage seepage zones. Favours extensive tracts of reeds and cane grass.

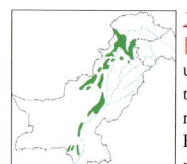

2 STRIATED PRINIA *Prinia criniger*   16 cm

[a] **ADULT BREEDING** and [b] **ADULT NON-BREEDING** Large size, with streaked upperparts, and stout bill. In breeding plumage, has dark bill and lores, and indistinct streaking to grey-brown upperparts. Non-breeding has prominently streaked rufous-brown upperparts, and buff lores. Song is a monotonous wheezy stridulation. Frequent and sedentary resident in warmer lower hilly areas, from Sind Kohistan north to Chitral, Swat, the Margallas and higher hills in the Punjab Salt Range. Associated with open scrub-covered slopes, Mazri Palm in the south and subtropical Pine in the north.

3 RUFOUS-FRONTED PRINIA *Prinia buchanani*   12 cm

ADULT Rufous-brown crown and broad white tips to tail (very prominent in flight). Rufous to crown may be difficult to see when worn. Upperparts uniform pale rufous-brown on juvenile. Cheery tinkling song is a *trich – a-tree-trich-a-tree*. Call is a distinctive rippling trill. Common resident, confined to scrub desert areas throughout the Indus basin, extending westwards along the Mekran coast, and also in Sind and the NWFP foothills; avoids only irrigated cropland, but adapted to irrigated forest plantations.

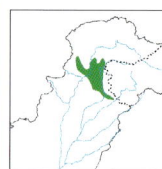

4 GREY-BREASTED PRINIA *Prinia hodgsonii*   11 cm

[a] **ADULT BREEDING** and [b] **ADULT NON-BREEDING** Small size. Diffuse grey breast-band in summer. In non-breeding plumage, has fine dark bill, fine whitish supercilium, brown upperparts with rufescent cast, and greyish wash to sides of neck and breast. Males perform wing-snapping display and tinkling song, *yousee-yousee-yousee*. Locally common resident, restricted to dense thorn scrub or dry deciduous scrub forest on southern slopes from Swat and Hazara District to the Margallas.

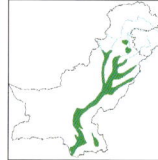

5 YELLOW-BELLIED PRINIA *Prinia flaviventris*   13 cm

ADULT White throat and breast, and yellow belly. Slate-grey cast to crown and olive-green cast to upperparts. Juvenile has uniform yellowish olive-brown upperparts and yellow underparts. Brief song is a descending flute-like *tree-dula-lu-lee*. Locally common resident along major riverine tracts, major irrigation seepage areas and lakes, wherever extensive reed-beds occur. Has benefited from irrigation schemes.

6 ASHY PRINIA *Prinia socialis*   13 cm

[a] **ADULT BREEDING** and [b] **ADULT NON-BREEDING** Slate-grey crown and ear-coverts, red eyes, slate-grey or rufous-brown upperparts, and orange-buff wash to underparts. Contact calls are a plaintive *chee-chee*; song comprises 2-noted chinking whistles, *tchiup-tchiup*. Locally common sedentary resident, confined to northeast Punjab, but gradually spreading southwestwards. Prefers cropland and urban gardens in well-wooded areas.

7 PLAIN PRINIA *Prinia inornata*   13 cm

[a] **ADULT BREEDING** and [b] **ADULT NON-BREEDING** In breeding plumage, has greyish upperparts, dark bill, and shortish tail with prominent white outertail feathers. More rufescent in non-breeding plumage, with longer tail. Song is a monotonous insect-like reeling, *dzwee-dzwee*. Common resident throughout the Indus basin, favouring tall irrigated crops, especially sugarcane, and in drier scrub desert near swampy or seepage areas. Unlike the Ashy Prinia, it shuns urban gardens or well-wooded areas. Some local summer migration takes place into adjacent foothills.

MISCELLANEOUS WARBLERS, PLATE 74

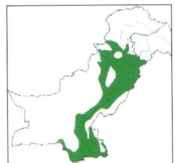

1 GRACEFUL PRINIA *Prinia gracilis* 11 cm
ADULT Small, with fine bill and streaked upperparts. Much smaller than Rufous-vented, with more uniform sandy grey-brown coloration to upperparts and uniform whitish underparts lacking rufous vent. Males perform vertical jerky wing-snapping display and insect-like reeling song, *dzit-dzit–dzit*. Common resident throughout riverine tracts from the Indus plains south to East Narra and north to Bannu and Mardan districts. Avoids dry scrub desert or intensively cultivated areas.

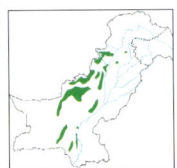

2 STREAKED SCRUB WARBLER *Scotocerca inquieta* 10 cm
ADULT Vinaceous-buff supercilium and ear-coverts, black throat streaking, grey-ish upperparts, and boldly streaked crown. Contact calls are a rapid descending trill, *twit-ti-tu-tu-tuh*. Song comprises high-pitched squeaking warbles ending in a drawn-out trill. Frequent resident throughout Baluchistan, NWFP, Sind Kohistan and the Kalla Chita Range. Adapted to dry open hillsides with *Artemisia* or scattered thorn bushes and dry juniper, or Chilghoza forest in northern Baluchistan up to 2500 m.

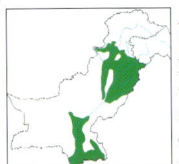

3 ZITTING CISTICOLA *Cisticola juncidis* 10 cm
a ADULT BREEDING and **b** ADULT NON-BREEDING Small, with a short tail that has prominent white tips. Bold streaking on buff upperparts, including nape, and thin whitish supercilium. Males perform a high circling display with spaced clicking calls. Common but erratically occurring resident, confined to open boggy or marshy grassland, avoiding tall reeds and seasonally adapted to flooded paddy and irrigated green crops, such as Egyptian Clover. Extends into the Kurram and Peshawar valleys. Rare in southern Punjab and northern Sind.

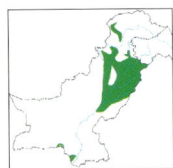

4 ORIENTAL WHITE-EYE *Zosterops palpebrosus* 10 cm
ADULT Black lores and white eye-ring, bright yellow throat and breast, and whitish belly. Contact calls are a querulous *cheer-cheer*. Song comprises rising and falling whistles followed by similar querulous 2-toned notes. Abundant arboreal resident in Punjab and northeastern NWFP, with a small population in mangroves in the Indus delta and Sonmeani, extending in summer into the outer Himalayas where deciduous forest predominates, up to 2400 m.

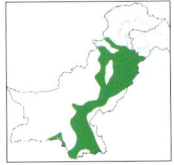

5 COMMON TAILORBIRD *Orthotomus sutorius* 13 cm
a MALE and **b** FEMALE Rufous forehead, greenish upperparts, and whitish underparts including undertail-coverts. Territorial song comprises a monotonous 2-noted *chwi-chwi*. Common resident throughout the Indus basin and western fringes of NWFP. Adapted to mangroves, dry scrub forest, urban gardens, irrigated forest plantations, and Himalayan foothills up to 900 m, shunning only extensive desert and dry hilly tracts.

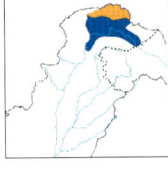

6 WHITE-BROWED TIT WARBLER *Leptopoecile sophiae* 10 cm
a MALE and **b** FEMALE Whitish supercilium, rufous crown, and lilac and purple in plumage. Frequent resident in far northern areas throughout the Karakorams and eastern Gilgit. Breeds above 3000 m, but descends in winter to valley bottoms where river banks have good bush cover (down to Besham, on the Indus, at 900 m). Not recorded from Chitral or Swat.

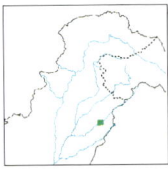

7 STRIATED GRASSBIRD *Megalurus palustris* 25 cm
a ADULT WORN and **b** ADULT FRESH Streaked upperparts, finely streaked breast, and long, graduated tail. Has longer and finer bill, and more prominent supercilium, than Bristled Grassbird (see Appendix). Song often accompanied by gradually descending flight display, and very loud, rich, fruity notes alternating with trills and drawn-out warbles: *chot-chot-chot-which-u-quieee-chot-trrrt-kwit-kwit – chwot*. Rare, with only one known isolated population on the Ravi river at Balloki Headworks.

SYLVIA WARBLERS, PLATE 75

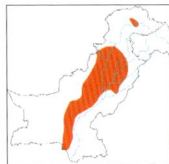

1 GREATER WHITETHROAT *Sylvia communis* 14 cm
a MALE and **b** FEMALE Larger and longer-tailed than Lesser, with broad, well-defined, sandy-brown to pale rufous-brown fringes to greater coverts and tertials, pale base to lower mandible, and orange-brown to pale brown (not grey) legs and feet. Uncommon early autumn migrant, with spring passage extra-limitally to the west. On passage, favours bush-dotted scrub jungle; especially numerous in Sind Kohistan just before crossing the Indian Ocean.

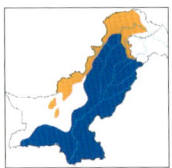

2 LESSER WHITETHROAT *Sylvia curruca* 13 cm
a ADULT *S. c. blythi*, **b** ADULT *S. c. minula* and **c** ADULT *S. c. althaea* Brownish-grey upperparts, grey crown with darker ear-coverts, blackish bill, and dark grey legs and feet. *S. c. althaea* ('Hume's Whitethroat') is the largest and darkest race; *minula* ('Small Whitethroat') is the most finely built, with pale, sandy grey-brown upperparts. Song is a loud, continuous, bubbling warble. *S. c. althaea* breeds commonly in the drier montane regions of western and northwestern Pakistan, whilst wintering mostly in peninsular India. *S. c. minula* winters in Pakistan, favouring the major desert tracts of Cholistan and Thar. *S. c. blythi*, which breeds in Siberia, is abundant throughout the winter in Sind, Punjab and western NWFP, especially in irrigated canal areas.

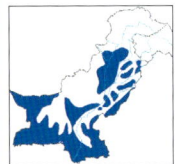

3 DESERT WARBLER *Sylvia nana* 11.5 cm
ADULT Small size, sandy-brown upperparts, rufous rump and tail, yellow iris, and yellowish legs and feet. Common winter visitor in open desert habitat in southern Baluchistan, eastern parts of NWFP, and Sind and Punjab, whether stony gravel plains or wind-blown sand dunes.

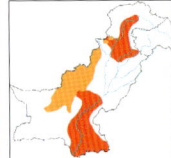

4 ORPHEAN WARBLER *Sylvia hortensis* 15 cm
a ADULT MALE, **b** ADULT FEMALE and **c** 1ST-WINTER Larger and bigger-billed than Lesser Whitethroat, with more ponderous movements and heavier appearance in flight. Adult has blackish crown, pale grey mantle, blackish tail and pale iris. First-year has crown concolorous with mantle, with darker grey ear-coverts and dark iris; very similar to many Lesser Whitethroats. Orphean often shows darker-looking uppertail, lacks or has indistinct eye-ring, and has greyish centres and pale fringes to undertail-coverts; these features are variable and difficult to observe in the field. Frequent double passage migrant through the Indus basin, with a small breeding population in higher bush-covered hills of northern Baluchistan and northwestern NWFP (760–2500 m). A few overwinter in southern Sind, where they favour riverine forest or dense scrub jungle.

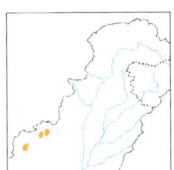

5 MÉNÉTRIES WARBLER *Sylvia mystacea* 12 cm
a MALE and **b** FEMALE Small size, with long tail. Male has blackish hood, reddish-brown eye-ring, and pinkish wash to throat and breast, with white sub-moustachial line. Female/first-winter has rather uniform sandy grey-brown upperparts (lacking darker mask), dark tail, and pale legs and feet. Like all members of the genus, it keeps up sharp *tak-tak* contact calls. Song comprises a long, continuous, bubbling, chattering warble, with some whistling squeaky notes. Summer visitor, and rare breeding bird in valley bottoms with good bush cover, west-central Baluchistan.

BUSH WARBLERS AND *ACROCEPHALUS* WARBLERS, PLATE 76

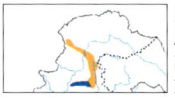

1 **BROWNISH-FLANKED BUSH WARBLER** *Cettia fortipes* 12 cm
ADULT Olive-brown upperparts and rather indistinct supercilium. Underparts are buffish-grey with brownish-olive flanks (Cetti's Warbler is whiter on underparts). Loud song, *p-e-e-e-e-h-swit-tsyiu*. Breeding on lower open Himalayan slopes, including terraced cultivation. Winters in adjacent foothills in reed-beds and bush-covered ravines.

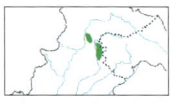

2 **GREY-SIDED BUSH WARBLER** *Cettia brunnifrons* 10 cm
ADULT Chestnut crown, whitish supercilium and greyish underparts. Song starts with 2 piercing whistles followed by *whicheeta-whicheeta*. Rare resident from Hazara District to the Neelum Valley, preferring sheltered slopes towards the upper limit of the tree-line (2900–3500 m). Winters in adjacent foothills down to 1200 m.

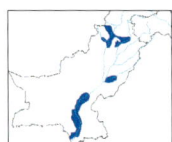

3 **CETTI'S BUSH WARBLER** *Cettia cetti* 14 cm
ADULT Large size, with big tail; white breast and indistinct supercilium, greyish breast sides and flanks, and whitish tips to undertail-coverts. Song is an explosive *chit-chit-chitity-chit – chitity-chit*. Uncommon but locally frequent winter visitor to southern Sind, and northeastern NWFP and Punjab, favouring dense reed-beds and tall rank vegetation around lakes and main rivers.

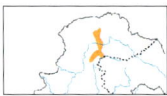

4 **LONG-BILLED BUSH WARBLER** *Bradypterus major* 13 cm
a b ADULT WORN and **c** ADULT FRESH Spotting on throat and breast (can be indistinct). Long bill, white underparts and unmarked undertail-coverts. Song is a monotonous *pikha-pikha pikha*. Uncommon breeding visitor to inner Himalayan valleys from Gilgit to northern Hazara District, favouring forest edges and terraced cultivation with good herbaceous or bush cover (2400–3500 m). Winters in adjacent foothills down to 1200 m.

5 **GRASSHOPPER WARBLER** *Locustella naevia* 13 cm
a b ADULT Olive-brown upperparts with dark streaking, indistinct supercilium, and unmarked or only lightly streaked underparts. Underparts have a yellowish wash in first-winter plumage. Scarce winter visitor, mainly recorded in northern parts of NWFP and Punjab. Swampy areas alongside lakes and seepage zones of major rivers. Spring passage mainly in Kohat District, autumn passage through Gilgit.

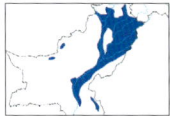

6 **MOUSTACHED WARBLER** *Acrocephalus melanopogon* 12.5 cm
ADULT Broad white supercilium, blackish eye-stripe and boldly streaked rufous-brown mantle. Often cocks tail in chat-like fashion. Very common winter visitor along the Indus and its tributaries, and larger lakes, wherever reed-beds or tamarisks subject to seasonal flooding occur. Possibly breeds Baluchistan.

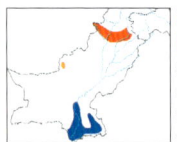

7 **PADDYFIELD WARBLER** *Acrocephalus agricola* 13 cm
a ADULT FRESH and **b** ADULT WORN Prominent white supercilium behind eye, and stout bill with dark tip. Often shows dark edge to supercilium. Rufous cast to upperparts in fresh plumage. (See also Table 6 on p.245.) Common winter visitor to lower Sind; a few breed in northwestern Baluchistan, with double passage migration across NWFP and northern Punjab. Prefers reed-beds and wet marshy ground, using paddy-fields and sugarcane on passage.

8 **BLUNT-WINGED WARBLER** *Acrocephalus concinens* 13 cm
a ADULT FRESH and **b** ADULT WORN Longer and stouter bill than Paddyfield, with pale lower mandible. Supercilium indistinct behind eye. (See also Table 6 on p.245.) Rare breeder in northern alpine valleys, Hazara District. No winter records.

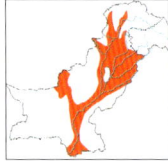

9 **BLYTH'S REED WARBLER** *Acrocephalus dumetorum* 14 cm
a ADULT FRESH and **b** ADULT WORN Long bill, olive-brown to olive-grey upperparts, and uniform wings. Supercilium indistinct compared with Paddyfield, and is barely apparent behind eye. (See also Table 6 on p.245, and see Appendix for comparison with Eurasian Reed.) Common double passage migrant from main valleys in northern areas, through central Baluchistan, down the Indus to the coast. Favours tree groves, good bush cover and shady gardens. Not associated with reeds.

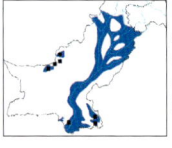

10 **CLAMOROUS REED WARBLER** *A. stentoreus* 19 cm
ADULT Large size, with long bill, short primary projection and whitish supercilium. (See Appendix for comparison with Great Reed.) Common winter visitor throughout the Indus basin, wherever lakes and rivers provide swampy conditions. Scattered breeding records from lower Sind, Baluchistan lakes and the Kurram Valley.

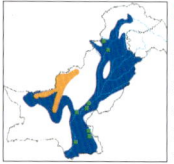

1 BOOTED WARBLER *Hippolais caligata* — 12 cm
a ADULT *H. c. caligata* and **b** ADULT *H. c. rama* Small size and *Phylloscopus*-like behaviour. Tail looks long and square-ended, and undertail-coverts look short. Often shows faint whitish edges and tip to tail and fringes to tertials. (See also Table 6 on p.245.) *H. c. caligata* is common double passage migrant through the Indus basin, with *H. c. rama* wintering on the Indus plains and breeding in western Baluchistan, with scattered breeding records from East Narra and along the Indus. Favours tree groves, nesting near water on the Indus plains and in orchards in Baluchistan.

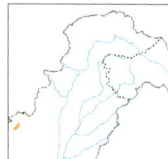

2 UPCHER'S WARBLER *Hippolais languida* — 14 cm
ADULT Large size and heavy appearance. Long, full tail is often swayed or flicked downwards. Grey upperparts, often with dark wings and tail. Tail usually shows white sides and tip; usually whitish fringes to wing-coverts and tertials. (See also Table 6 on p.245.) Uncommon summer breeding visitor to west-central Baluchistan, on open *Artemisia*-covered hillsides and ravines with scattered bushes (1800–2400 m).

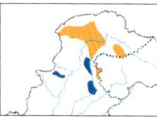

3 GOLDCREST *Regulus regulus* — 9 cm
ADULT Small size. Plain face, lacking a supercilium, but with a dark eye and pale eye-ring. Yellow centre to crown. Uncommon resident, breeding in Himalayan coniferous forest (2500–3300 m) from Chitral down to the Murree Hills, with limited altitudinal movement in winter.

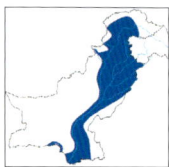

4 COMMON CHIFFCHAFF *Phylloscopus collybita* — 11 cm
ADULT Brownish to greyish upperparts; olive-green edges to wing-coverts, remiges and rectrices. Black bill and legs. No wing-bar. (See also Table 2 on p.242.) Abundant winter visitor throughout the Indus plains and northern area valleys, September to May. Males start singing from mid-March while on passage.

5 MOUNTAIN CHIFFCHAFF *Phylloscopus sindianus* — 11 cm
ADULT Much as Common Chiffchaff, but generally lacks olive-green tone to back and to edges of wing-coverts, remiges and rectrices. Different call. (See also Table 2 on p.242.) Breeds in bushes; winters in riverine trees and bushes. Frequent resident, breeding in inner drier valleys from Chitral to Baltistan, favouring open bush-dotted slopes, where it is sympatric with Tickell's Leaf Warbler (2300–3600 m). Winters mainly in Sind, from Jacobabad to Nawabshah and East Narra.

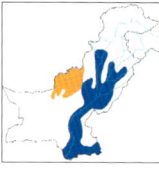

6 PLAIN LEAF WARBLER *Phylloscopus neglectus* — 10 cm
ADULT Much as Common Chiffchaff, but smaller, with shorter tail, no olive-green or yellow in plumage, and hard *tak-tak* call. Short song is a *chit-chuwich-chissa-chit-chu-twissa twissa-twit*. (See also Table 2 on p.242.) Common resident, breeding in northern Baluchistan in juniper and Chilghoza forest, above 2400 m. Winters in damper well-wooded areas, along the Indus and its tributaries. More plentiful in Sind than Punjab.

7 TICKELL'S LEAF WARBLER *Phylloscopus affinis* — 11 cm
ADULT Dark greenish to greenish-brown upperparts, and bright yellow supercilium and underparts. *Chit* call. Brief song is a *chit-teiu-teiu-teiu-teiu*. (See also Table 2 on p.242.) Common summer visitor, breeding at 3200–4200 m in alpine or subalpine regions of the far north from Chitral to Baltistan. Absent from Safed Koh and the Murree Hills. Winters in the Indian peninsula and northeastern Himalayan foothills.

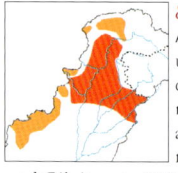

8 SULPHUR-BELLIED WARBLER *Phylloscopus griseolus* — 11 cm
ADULT Dark greyish upperparts, bright yellow supercilium, and dusky yellow underparts strongly washed with buff. (See also Table 2 on p.242.) Distinctive contact call is a loud, hard *quit-quit*. Song starts with thin whistles followed by rapid repetition: *tseep-tyi-tyi-tyi-tyi*. Has distinctive habit of climbing about rocks and nuthatch-like on tree trunks and branches. Common breeding bird in drier rock-strewn slopes of northern Baluchistan above 2400 m, and Safed Koh, Chitral and Gilgit up to 3500 m. Winters in the Indian peninsula, with marked spring passage to the plains of NWFP and northern Punjab.

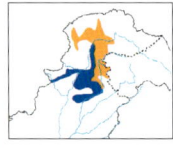

1 LEMON-RUMPED WARBLER *Phylloscopus chloronotus* 9 cm

ADULT Yellowish crown-stripe and rump-band, dark crown-sides, dark panel across greater coverts, and whitish underparts. (See also Table 4 on p.243.) Call is a high-pitched *uist*. Frequent resident, breeding in Himalayan coniferous forest from Swat east to the Neelum Valley and Murree Hills. Prefers damper northern slopes with Silver Fir and spruce (2200–2700 m). Winters in mountain valleys.

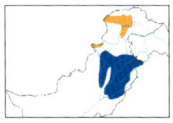

2 BROOKS'S LEAF WARBLER *Phylloscopus subviridis* 9 cm

a ADULT FRESH and **b** ADULT WORN Yellow supercilium and ear-coverts, indistinct yellow crown-stripe, variable yellow wash to throat and breast, and indistinct pale rump-band. (See also Table 4 on p.243.) Call is a loud and piercing *chwee*. Frequent resident in Himalayan coniferous forest in inner cooler valleys, from Chitral east through Indus Kohistan to forested valleys of southern Gilgit. Winters from the foothills of NWFP across Punjab, south to Bahawalpur.

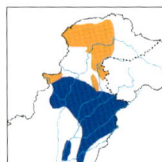

3 HUME'S WARBLER *Phylloscopus humei* 10–11 cm

ADULT Lacks rump-band and well-defined crown-stripe. Has buffish or whitish wing-bars and supercilium. Bill appears all dark, and legs are normally blackish-brown. (See also Table 4 on p.243.) Call is a rolling *whit-hoo*. Brief song is a repeated *tissoo-tissoo* interspersed with a buzzing *z-z-z-zyip*. Very common, breeding in Himalayan forest from Safed Koh and throughout northern areas up to subalpine birch scrub (2500–3500 m). Winters in better wooded parts of NWFP and Punjab.

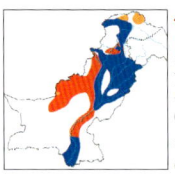

4 GREENISH WARBLER *Phylloscopus trochiloides* 10–11 cm

a ADULT *P. t. viridanus* and **b** ADULT *P. t. nitidus* Prominent supercilium, but lacking crown-stripe. Has fine wing-bar (sometimes 2). Throat and breast have yellow wash in *P. t. nitidus*. Slurred, loud *chli-wee* call is quite different from Large-billed. (See also Table 3 on p.243.) *P. t. viridanus* is common double passage migrant through northern Punjab and Vale of Peshawar. Scattered breeding in subalpine birch scrub on precipitous slopes (2700–4200 m). *P. t. nitidus* follows a more westerly passage through Baluchistan, wintering in southern Sind and occasional breeding in northwestern Baluchistan.

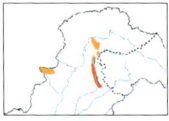

5 LARGE-BILLED LEAF WARBLER *Phylloscopus magnirostris* 13 cm

ADULT Clear, loud *der-tee* call is best feature from Greenish. Large, with large dark bill, bold yellow-white supercilium and dark eye-stripe. (See also Table 3 on p.243.) Stereotyped song has 5 spaced whistles, *tee-ti-tiii-tu-tuh*. Scarce summer breeder of mountain stream-banks in Himalayan temperate forest (2100–3500 m). Winters in South India.

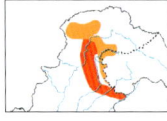

6 TYTLER'S LEAF WARBLER *Phylloscopus tytleri* 11 cm

ADULT Slender, mainly dark bill, long fine supercilium, no wing-bars and shortish tail. (See also Table 2 on p.242.) Short song, with 4 rising and falling sibilant phrases, *kichu-qwishu-kitchu-qwishu*. Breeds in Swat forested slopes, from Swat east to Gilgit, northern Hazara District and the Neelum Valley. Winters extra-limitally.

7 WESTERN CROWNED WARBLER *Phylloscopus occipitalis* 11 cm

ADULT Crown-stripe, large size, greyish-green upperparts and greyish-white underparts. While hunting, calls *chiwee-chiwee* and flicks one wing half-open constantly. (See also Table 5 on p.244, and see Appendix for differences from Blyth's Leaf.) The most abundant leaf warbler in Pakistan, breeding in Himalayan forest from Chitral to the Neelum Valley and Murree Hills (2000–3200 m). Winters in peninsular India.

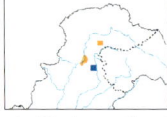

8 WHISTLER'S WARBLER *Seicercus whistleri* 10 cm

ADULT Green upperparts, yellow underparts, dark sides to crown and yellow eye-ring. Now recognised as a distinct species from Golden-spectacled *S. burkii*, which is not known from Pakistan. Song lacks tremelos and trills of that species. Very rare resident, possibly breeding at upper limit of tree-line in Hazara District and the Neelum valley. Winters in outer foothills.

9 GREY-HOODED WARBLER *Seicercus xanthoschistos* 10 cm

ADULT *Phylloscopus*-like appearance. Whitish supercilium, and grey crown with pale central stripe; yellow underparts and no wing-bars. Song is a *tsi-weetsi-weetsi – weetu-ti-tu*. Locally common resident in Himalayan foothills and lower warmer valleys (900–2100 m). Wanders to adjacent plains in winter. Not recorded west of Swat.

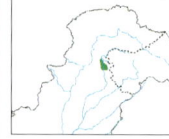

1 **WHITE-THROATED LAUGHINGTHRUSH**
Garrulax albogularis 28 cm
ADULT White throat and upper breast, rufous-orange belly, and broad white tip to tail. Contact call is a cat-like *teer-teer*. Rare gregarious resident, surviving in lower watershed of the Jhelum river. Found in Dunga Gali up until the 1960s, now confined to Poonch. Prefers predominantly deciduous Himalayan forest 1500–2440 m.

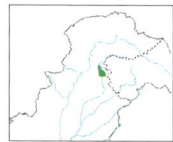

2 **RUFOUS-CHINNED LAUGHINGTHRUSH**
Garrulax rufogularis 22 cm
ADULT Blackish cap, rufous chin and throat, black barring on upperparts, bold patterning on wing, and rufous-orange tip and black subterminal band to tail. Very rare resident, occurring only in the Jhelum Valley, Poonch, at lower altitudes with dense bush cover (600–1800 m).

3 **STREAKED LAUGHINGTHRUSH** *Garrulax lineatus* 20 cm
ADULT Small, and mainly grey-brown with fine pale streaking. Grey-tipped tail, and grey panel in wings. Foraging groups utter 4 loud plaintive whistles in descending scale. Territorial song comprises loud trisyllabic whistles, *pitt-wee-yarr*. Common resident, well adapted to arid cold mountain regions, extending from higher scrub-covered hills in Quetta District north through NWFP, the Kurram Valley and the whole of northern areas, as well as the Murree Hills (2000–3000 m). Wanders to foothills and main valley floors in winter (1200 m).

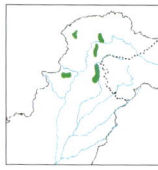

4 **VARIEGATED LAUGHINGTHRUSH**
Garrulax variegatus 24 cm
ADULT Black eye-patch and throat, rufous-buff forehead, black patches on greyish wings, and grey tip to tail. Warning calls are grating squeaks, *gweek-gweek*. Territorial song comprises ringing whistles, *pit-we-weeer*. Common resident in Himalayan forest, restricted to moister slopes and ravines with good bush cover (1500–3000 m). Rare in lower Chitral and Swat, more abundant in Hazara District and the Murree Hills. Descends in winter to 600 m.

5 **RUSTY-CHEEKED SCIMITAR BABBLER**
Pomatorhinus erythrogenys 25 cm
ADULT Rufous sides of head and neck, and rufous sides of breast and flanks contrasting with white of rest of underparts. Pairs duet at dawn and dusk; males with a loud ringing *khwoi*, females instantly responding with a staccato *quip*. Locally frequent sedentary resident, confined to outer foothills of the Himalayas (450–2100 m). Prefers dense bush-covered hillsides and dry subtropical deciduous forest.

6 **BLACK-CHINNED BABBLER** *Stachyris pyrrhops* 10 cm
ADULT Black chin and lores, and orange-buff underparts. Song is an accelerating piping, like tiny bells. Locally restricted, frequent resident from the Margallas to eastern spurs of the Murree Hills and the Neelum Valley, favouring lower elevations with thick bush cover, especially shady ravines (450–1500 m).

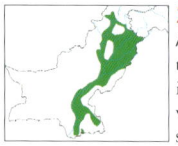

7 **YELLOW-EYED BABBLER** *Chrysomma sinense* 18 cm
ADULT Yellow iris and orange eye-ring, white lores and supercilium, and white throat and breast. Song comprises rapid twittering trills. Frequent sedentary resident throughout the Indus basin, though patchily distributed, preferring the vicinity of rivers and tall cane grass mixed with thorn bushes, but adapted to shady suburban gardens and irrigated forest plantations in northern Punjab.

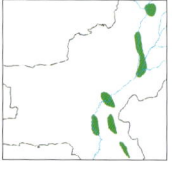

8 **JERDON'S BABBLER** *Chrysomma altirostre* 17 cm
ADULT Brown iris, grey lores and supercilium, and grey throat and breast. Territorial calls are a plaintive *teuw-teuw*. Song lacks the rapid trills of that of the Yellow-eyed, comprising rising and falling 2-toned whistles. Very rare resident in 3 disjunct riverine populations in Muzafargarh, Khairpur and Sangar districts. Globally threatened (Vulnerable).

BABBLERS, PLATE 80

1 **COMMON BABBLER** *Turdoides caudatus*   23 cm
ADULT Streaked upperparts. Unstreaked whitish throat and breast centre. Legs and feet are yellowish. Family groups when going to roost indulge in territorial choruses, a series of rapid trilling whistles. Abundant resident throughout the Indus basin, extending eastwards through major desert tracts and westwards into dry foothill areas (sea-level–1500 m). A small separate population, *T. c. huttoni*, occurs in west-central border regions of Baluchistan in open tamarisk tracts. Avoids closed canopy woodland and mountainous regions.

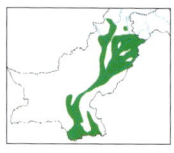

2 **STRIATED BABBLER** *Turdoides earlei*   21 cm
ADULT Streaked upperparts. Brown mottling on fulvous throat and breast. Legs and feet are greyish to olive-brown. Common sedentary resident throughout the Indus plains, preferring riverine tracts with reed-beds or dense cane grass, but has colonised major irrigation systems and main rice-growing tracts. Waterlogging of cultivated areas has favoured this species.

3 **LARGE GREY BABBLER** *Turdoides malcolmi*   28 cm
ADULT Dull white sides to long, graduated tail; unmottled pinkish-grey throat and breast, pale grey forehead and dark grey lores. Rare and erratic visitor to north-east border regions of Punjab, from Lahore to Kasur, favouring mango orchards and irrigated forest plantations.

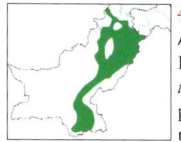

4 **JUNGLE BABBLER** *Turdoides striatus*   25 cm
ADULT Uniform tail; variable dark mottling and streaking on throat and breast. Family groups frequently indulge in noisy choruses of plaintive squeaks and *tew-tew* calls. Abundant territorial and sedentary resident, throughout better wooded parts of the Indus basin, avoiding only extensive desert or dry foothill regions (up to 1800 m).

5 **RED-BILLED LEIOTHRIX** *Leiothrix lutea*   13 cm
MALE Red bill, yellowish-olive crown, yellow throat and orange breast, and forked black tail. Male has some crimson edgings to flight feathers; in female, the crimson is replaced by yellow. Irregular winter visitor to the Margalla Hills and foothills of the southeast Murree Hill range. Often in parties, remaining several weeks in the same area and hunting in dense bushy undergrowth (up to 2000 m).

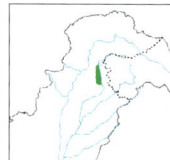

6 **WHITE-BROWED SHRIKE BABBLER**
   Pteruthius flaviscapis   16 cm
a MALE and **b** FEMALE Male has rufous tertials, black cap with white supercilium, and grey mantle. Female has grey cap and olive mantle; its larger size and rufous tertials best distinguish it from Green. Warning calls are a grating *chrri* and contact calls are trisyllabic pleasant whistles, *yup-yipy-up*. Scarce resident confined to the Murree Hill range at lower altitudes where 'Ban' oak and deciduous trees predominate (900–2000 m).

7 **GREEN SHRIKE BABBLER** *Pteruthius xanthochlorus*   13 cm
MALE Grey crown and nape, narrow white or yellowish-white wing-bar, greyish-white throat and breast, and yellowish flanks. Song is a rapid repetition of single piping notes. Very rare resident, largely confined to the Neelum Valley, and no records from the Murree Hills since 1900. Inhabits Himalayan moist forest with mixed conifers and deciduous trees (2000–3000 m).

8 **RUFOUS SIBIA** *Heterophasia capistrata*   21 cm
ADULT Black cap, and rufous collar and underparts; black and grey bands on rufous tail. Song is a loud piping whistle, *tee-dee-dee-deeeh*. Locally frequent resident, confined to moist mixed deciduous/coniferous forest in the Murree Hill range, ascending in summer to 2700 m and in winter to 460 m. Omnivorous and largely arboreal in its habits.

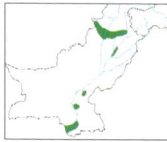

1 SINGING BUSHLARK *Mirafra cantillans*   14 cm
ADULT Stout bill, and rufous on wing; white outer tail feathers, limited spotting on breast, and whitish throat with buff breast-band. Song, delivered in flight, is sweet and full, with much mimicry. Scarce resident and local migrant, inhabiting semiarid open country and avoiding sandy desert. Breeds Salt Range plateau and fallow fields in lower Sind. Occasional straggler across Punjab plains.

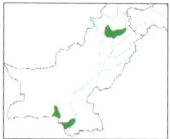

2 INDIAN BUSHLARK *Mirafra erythroptera*   14 cm
ADULT Stout bill, and rufous on wing (more prominent than on Singing); rufous-buff on outer tail feathers, dark spotting on breast and more uniform whitish underparts compared with Singing. Song is a *tit-tit-tit*, followed by drawn-out *tsweeeih-tsweeeih-tsweeeih*. Resident from Las Bela across lower Sind to Thar Desert. Winter visitor to plateau of Salt Range. Inhabits low hilly areas and barren clay flats.

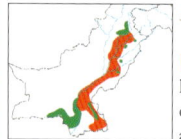

3 BLACK-CROWNED SPARROW LARK
Eremopterix nigriceps   13 cm

a MALE and **b** **c** FEMALE Male has blackish crown and underparts. Female has stout bill, rather uniform head and upperparts, and dark grey underwing-coverts. Song during 'parachute' display flight is *dwee-di-ul-twee-e-h*, followed by a prolonged mournful whistle. Locally common nomadic resident throughout desert tracts and dry valleys in southern Baluchistan and Punjab Salt Range. Nests semi-colonially.

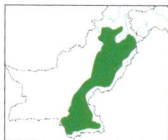

4 ASHY-CROWNED SPARROW LARK
Eremopterix grisea   12 cm

MALE Male has grey crown and nape and blackish underparts. Female inseparable from Black-crowned. Short song during display flights is *tweedle –deedle-dee-dle*. Locally common but nomadic resident throughout the Indus basin, preferring uncultivated tracts but avoiding extensive sand-dune desert.

5 BAR-TAILED LARK *Ammomanes cincturus*   15 cm
ADULT Blackish terminal bar to rufous tail; tertials are orangey buff and contrast with blackish primaries. Smaller and stockier than Desert, with shorter and finer bill. Rather weak fluty song during dipping display flights. Common resident in southern Baluchistan only, inhabiting low stony foothills and bleak gavelly plains.

6 RUFOUS-TAILED LARK *Ammomanes phoenicurus*   16 cm
ADULT Dusky grey-brown upperparts, rufous-orange underparts, prominent dark streaking on throat and breast, and rufous-orange uppertail-coverts and tail, with dark terminal bar. Song similar to that of Desert Lark, comprising flute-like *tee-whoo* interspersed with low-pitched whistles and chirrups. Uncommon monsoon visitor to plateau region of Punjab Salt Range. Prefers fallow fields, and open treeless areas with scattered bushes.

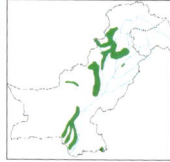

7 DESERT LARK *Ammomanes deserti*   17 cm
ADULT Tail largely grey-brown with rufous at sides. Told from Rufous-tailed by much paler upperparts, buffish to greyish-white underparts (with weak breast streaking) and lack of dark bar at tip of tail. Flight song is rather stereotyped, with spaced flute-like whistles, *peo-pyooh – peef-poof*, interspersed with warbling phrases. Common sedentary resident from sea coast to NWFP and lower Swat. Inhabits foothills, always in rocky or stony places, avoiding higher steeper slopes (760–1800 m).

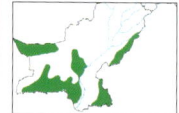

8 GREATER HOOPOE LARK *Alaemon alaudipes*   19 cm
a **b** ADULT Longish down-curved bill and long legs. Black-and-white wing patterning prominent in flight. Song comprises 3 or 4 sets of piping whistles, the last when completing display flight accelerando. Scarce resident wherever extensive tracts of desert occur, from Chaghai to Sibi, Thar and Cholistan.

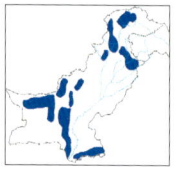

9 BIMACULATED LARK *Melanocorypha bimaculata*   17 cm
a **b** ADULT Large, with stout bill and short tail. Prominent white supercilium, black patch on side of breast, and white tip to tail. Frequent but erratic winter visitor throughout low flat plains or open valleys in Baluchistan, Tharparkar and northeast Punjab, and westwards to lower Chitral. A popular cage-bird in Pakistan due to its rich and powerful song.

LARKS, PLATE 82

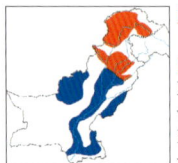

1 GREATER SHORT-TOED LARK
Calandrella brachydactyla 14 cm

a ADULT *C. b. dukhunensis* and **b** ADULT *C. b. longipennis* Stouter bill than Hume's, with more prominent supercilium and eye-stripe; upperparts generally warmer, with more prominent streaking; dark breast-side patches often apparent. Breast of *dukhunensis* has warm rufous-buff wash, especially at sides. Abundant winter visitor in flocks in fallow fields and scrub desert in Sind, Punjab and central Baluchistsan. Strong spring passage in Kurram Valley, Chitral and Gilgit.

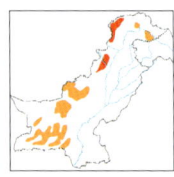

2 HUME'S SHORT-TOED LARK *Calandrella acutirostris* 14 cm
ADULT Greyer and less heavily streaked upperparts than Greater, with pinkish uppertail-coverts; dark breast-side patch usually apparent, with greyish-buff breast-band. Head pattern less pronounced than Greater's, with rather uniform ear-coverts. Uncommon breeding resident on lower *Artemisia* slopes in Baluchistan, NWFP and northern Baltistan (Shyok Valley). Avoids stony plateaus and higher peaks. Part of population winters in the plains, mainly in India.

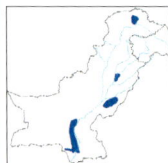

3 LESSER SHORT-TOED LARK *Calandrella rufescens* 13 cm
ADULT Primaries extend beyond tertials (tertials reach or almost reach tips of primaries in Greater and Hume's). Bill large, short and stout. Broad gorget of fine streaking on breast, fine but clear streaking on upperparts, streaked ear-coverts; whitish supercilia appear to join across bill. Scarce winter visitor, never encountered in flocks, Sind Kohistan, Bahawalnagar and Jhang districts, preferring arid foothill country with bare stony slopes.

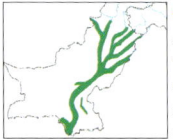

4 SAND LARK *Calandrella raytal* 12 cm
ADULT Small size, short tail, cold sandy-grey upperparts, and whitish underparts with fine sparse streaking on breast. Wing-tips clearly visible beyond tertials at rest; bill finer than in similar species. Swooping display flight accompanied by short song of continuous quavering warbles. Common sedentary resident, always in sand dunes close to rivers, from the sea coast to northern Hazara District.

5 CRESTED LARK *Galerida cristata* 18 cm
ADULT Large size and very prominent crest. Sandy upperparts and well-streaked breast; broad, rounded wings, rufous-buff underwing-coverts and outer tail feathers. Soaring display flight often at great heights, includes mimicry of other birds. The most widespread and abundant lark in Pakistan. Resident throughout Baluchistan, Punjab and Sind, with some winter migrant flocks on the Indus plains. Avoids irrigated cropland and steep mountainous regions.

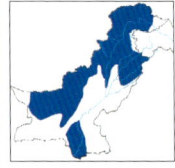

6 EURASIAN SKYLARK *Alauda arvensis* 18 cm
ADULT Large, with long tail. Pronounced whitish trailing edge to secondaries. Underparts whiter than on Oriental, has white outer tail feathers, lacks rufous wing-panel, and primaries extend noticeably beyond tertials (tertials almost reach primary tips in Oriental). Common winter visitor throughout lower Sind, western Baluchistan, northern Punjab and all northern areas. Prefers stubbles and grassy areas, avoiding stony or rocky terrain.

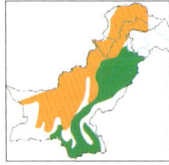

7 ORIENTAL SKYLARK *Alauda gulgula* 16 cm
ADULT Fine bill, buffish-white outer tail feathers, and indistinct rufous wing-panel. Abundant resident throughout the Indus basin and summer breeder from Baluchistan to the northern frontiers. Favours grassy places or short green crops, avoiding desert tracts. In northern areas, it breeds in high alpine regions where there are wide, flat valley bottoms. A fine songster and clever mimic.

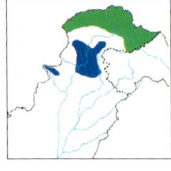

8 HORNED LARK *Eremophila alpestris* 18 cm
a MALE and **b** FEMALE Black-and-white head pattern and black breast-band. Common resident in far northern mountains, breeding from Chitral east to Baltistan in well-vegetated alpine tracts (3300–5100 m). Winters in adjacent terraced cultivation in foothills or in valley bottoms further north.

FLOWERPECKER, SUNBIRDS AND SPARROWS, PLATE 83

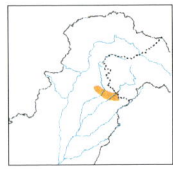

1 THICK-BILLED FLOWERPECKER *Dicaeum agile*   10 cm
ADULT The only flowerpecker recorded in Pakistan. Has thick bill, diffuse malar stripe, streaking on breast, and indistinct white tip to tail. Rare resident, confined to extreme southeastern foothills of the Murree Hills, inhabiting ravines with pistachio and sub-tropical pine and thick bush understorey (1000–1200 m). Winter visitor to adjacent plains in Sialkot and Gujrat districts.

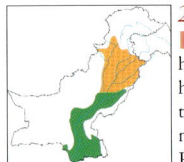

2 PURPLE SUNBIRD *Nectarinia asiatica*   10 cm
a MALE, **b** **c** FEMALE and **d** ECLIPSE MALE Male is metallic purple. Female has uniform yellowish underparts, with faint supercilium and darker mask; can have greyer upperparts and whiter underparts. Eclipse plumage male is similar to female but with dark stripe down centre of throat. Common resident, locally nomadic when nectar sources abundant, and migratory in summer to northern Punjab and Himalayan foothills (1200 m). Absent from dry mountainous regions; penetrates deserts when False Caper is in bloom.

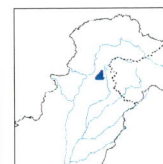

3 MRS GOULD'S SUNBIRD *Aethopyga gouldiae*   10 cm
a MALE and **b** FEMALE Male has metallic purplish-blue crown, ear-coverts and throat, crimson mantle and back (reaching yellow rump), yellow belly and blue tail. Female has pale yellow rump, yellow belly, short bill, and prominent white on tail. Rare and irregular winter visitor to the Margallas and Islamabad gardens, attracted to flowering creepers, especially *Loranthus*.

4 CRIMSON SUNBIRD *Aethopyga siparaja*   11 cm
a MALE and **b** FEMALE Male has crimson mantle, scarlet throat and breast, and grey belly. Female lacks yellow rump; is duller on underparts, has only indistinct pale tips to outer tail feathers, and has longer bill compared with female Mrs Gould's. Rare and irregular winter visitor to the Margallas and Islamabad gardens, attracted to flowering shrubs and trees (300–500 m).

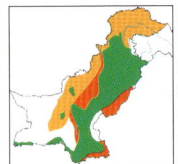

5 HOUSE SPARROW *Passer domesticus*   15 cm
a MALE and **b** FEMALE Male has grey crown, black throat and upper breast, chestnut nape and brownish mantle. Female has buffish supercilium and unstreaked greyish-white underparts. Very abundant sedentary resident *P. d. indicus* is locally migratory in summer in western Baluchistan, north to Chitral and Gilgit, always in association with human dwellings. Reaches higher densities in farming settlements, causing significant damage to ripening wheat. The migratory *P. d. bactrianus* is an abundant double passage migrant throughout the Indus plains, northern Baluchistan and NWFP, and a summer breeder colonially in far northern mountain valleys up to 3000 m.

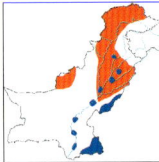

6 SPANISH SPARROW *Passer hispaniolensis*   15.5 cm
a MALE BREEDING, **b** MALE NON-BREEDING and **c** FEMALE Male has chestnut crown, black breast and streaking on flanks, and blackish mantle with pale 'braces'; pattern obscured by pale fringes in non-breeding season. Female told from female House Sparrow by longer whitish supercilium, fine streaking on underparts, and pale 'braces'. Common double passage migrant from central Asia, through northern mountains and central Baluchistan. Small numbers overwinter in major river-barrage seepage zones and main desert tracts in Sind and Punjab.

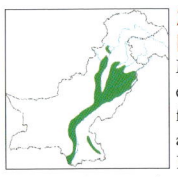

7 SIND SPARROW *Passer pyrrhonotus*   13 cm
a MALE and **b** FEMALE Smaller and slimmer than House, with finer bill. Male told from male House by restriction of chestnut on head to crescent around ear-coverts; also grey ear-coverts, and small black throat-patch. Female told from female House by more prominent buffish-white supercilium, greyish ear-coverts, and warmer buffish-brown lower back and rump. Frequent resident in Sind and Punjab in restricted habitat. Confined to major irrigation canals and riverine tracts with thick tamarisk subject to seasonal flooding. Spread of irrigation has favoured this species.

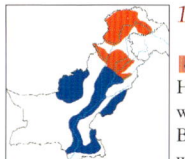

1 GREATER SHORT-TOED LARK
Calandrella brachydactyla   14 cm

a ADULT *C. b. dukhunensis* and **b** ADULT *C. b. longipennis* Stouter bill than Hume's, with more prominent supercilium and eye-stripe; upperparts generally warmer, with more prominent streaking; dark breast-side patches often apparent. Breast of *dukhunensis* has warm rufous-buff wash, especially at sides. Abundant winter visitor in flocks in fallow fields and scrub desert in Sind, Punjab and central Baluchistsan. Strong spring passage in Kurram Valley, Chitral and Gilgit.

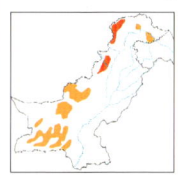

2 HUME'S SHORT-TOED LARK *Calandrella acutirostris*   14 cm
ADULT Greyer and less heavily streaked upperparts than Greater, with pinkish uppertail-coverts; dark breast-side patch usually apparent, with greyish-buff breast-band. Head pattern less pronounced than Greater's, with rather uniform ear-coverts. Uncommon breeding resident on lower *Artemisia* slopes in Baluchistan, NWFP and northern Baltistan (Shyok Valley). Avoids stony plateaux and higher peaks. Part of population winters in the plains, mainly in India.

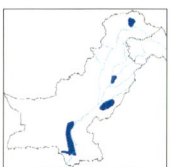

3 LESSER SHORT-TOED LARK *Calandrella rufescens*   13 cm
ADULT Primaries extend beyond tertials (tertials reach or almost reach tips of primaries in Greater's and Hume's). Bill large, short and stout. Broad gorget of fine streaking on breast, fine but clear streaking on upperparts, streaked ear-coverts; whitish supercilia appear to join across bill. Scarce winter visitor, never encountered in flocks, Sind Kohistan, Bahawalnagar and Jhang districts, preferring arid foothill country with bare stony slopes.

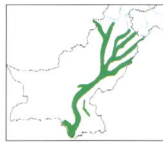

4 SAND LARK *Calandrella raytal*   12 cm
ADULT Small size, short tail, cold sandy-grey upperparts, and whitish underparts with fine sparse streaking on breast. Wing-tips clearly visible beyond tertials at rest; bill finer than in similar species. Swooping display flight accompanied by short song of continuous quavering warbles. Common sedentary resident, always in sand dunes close to rivers, from the sea coast to northern Hazara District.

5 CRESTED LARK *Galerida cristata*   18 cm
ADULT Large size and very prominent crest. Sandy upperparts and well-streaked breast; broad, rounded wings, rufous-buff underwing-coverts and outer tail feathers. Soaring display flight often at great heights, includes mimicry of other birds. The most widespread and abundant lark in Pakistan. Resident throughout Baluchistan, Punjab and Sind, with some winter migrant flocks on the Indus plains. Avoids irrigated cropland and steep mountainous regions.

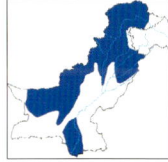

6 EURASIAN SKYLARK *Alauda arvensis*   18 cm
ADULT Large, with long tail. Pronounced whitish trailing edge to secondaries. Underparts whiter than on Oriental, has white outer tail feathers, lacks rufous wing-panel, and primaries extend noticeably beyond tertials (tertials almost reach primary tips in Oriental). Common winter visitor throughout lower Sind, western Baluchistan, northern Punjab and all northern areas. Prefers stubbles and grassy areas, avoiding stony or rocky terrain.

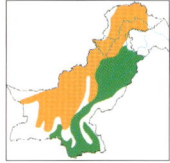

7 ORIENTAL SKYLARK *Alauda gulgula*   16 cm
ADULT Fine bill, buffish-white outer tail feathers, and indistinct rufous wing-panel. Abundant resident throughout the Indus basin and summer breeder from Baluchistan to the northern frontiers. Favours grassy places or short green crops, avoiding desert tracts. In northern areas, it breeds in high alpine regions where there are wide, flat valley bottoms. A fine songster and clever mimic.

8 HORNED LARK *Eremophila alpestris*   18 cm
a MALE and **b** FEMALE Black-and-white head pattern and black breast-band. Common resident in far northern mountains, breeding from Chitral east to Baltistan in well-vegetated alpine tracts (3300–5100 m). Winters in adjacent terraced cultivation in foothills or in valley bottoms further north.

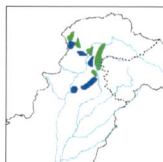

1 **RUSSET SPARROW** *Passer rutilans* 14.5 cm
a MALE and **b** FEMALE Male lacks black cheek-patch; has bright chestnut mantle and yellowish wash to underparts. Female has prominent supercilium and dark eye-stripe, rufous-brown scapulars and rump, and yellowish wash to underparts. Song more variable than that of House Sparrow: *chreet-chreet-chu-swik-chreet-chip*. Open forest, forest edges and cultivation. Locally common resident in forested areas of the Himalayas, breeding up to 2700 m, and descending in winter to adjacent valleys and foothills and better wooded areas of the Punjab Salt Range. Not recorded from Safed Koh, but occurs from Chitral southeastwards through Dir, Swat, Indus Kohistan and the Murree Hills. Replaced by Eurasian Tree Sparrow in the drier inner mountains.

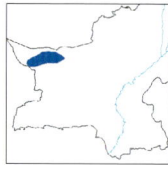

2 **DEAD SEA SPARROW** *Passer moabiticus* 12.25 cm
a MALE and **b** FEMALE Male has grey head with prominent buff supercilium and white submoustachial stripe; wing-coverts largely chestnut, and has yellow wash to underparts. Female has indistinct yellowish patch on neck, sandy-buff upperparts and yellow wash to underparts. Uncommon, locally restricted winter visitor (perhaps resident) to extreme northwestern Baluchistan in the Chaghai Desert. Bushes near non-perennial streams.

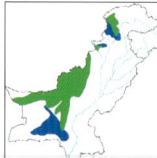

3 **EURASIAN TREE SPARROW** *Passer montanus* 14 cm
ADULT Chestnut crown, and black spot on ear-coverts. Sexes similar. Habitation and nearby cultivation. Locally common sedentary resident in drier western parts of Baluchistan, NWFP and lower Chitral, spreading to northwestern Swat and Gilgit in the Ghizar Valley (900–3000 m). Further east it is replaced by Russet Sparrow in Hazara District, and there is evidence that it is retreating further northwards with the spread of the more aggressive House Sparrow.

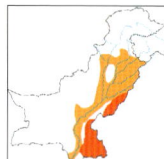

4 **CHESTNUT-SHOULDERED PETRONIA**
Petronia xanthocollis 13.5 cm
a MALE and **b** FEMALE Unstreaked brownish-grey head and upperparts, and prominent wing-bars. Male and some females have yellow on throat. Male has chestnut lesser coverts and white wing-bars; lesser coverts brown and wing-bars buff in female. Persistent singer, with monotonous *chulp-cheep-chillup-chip-cheep* etc. Mainly summer-breeding migrant throughout the Indus basin up to the outer Himalayan foothills, with a small population resident in lower Sind and on the Mekran coast. Favours riverine forest, irrigated forest plantations and well-wooded areas. Nests in natural tree cavities or building crevices.

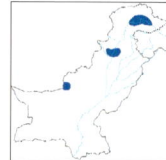

5 **ROCK SPARROW** *Petronia petronia* 17 cm
ADULT Striking head pattern, including pale crown-stripe. Has whitish patch at base of primaries and tip to tail. Rare winter visitor to northwest Baluchistan. More frequent in Kohat and Attock districts and southern Gilgit. Prefers dry stony ground or bare stubbles for winter foraging (250–1500 m).

1 FOREST WAGTAIL *Dendronanthus indicus* 18 cm
ADULT Broad yellowish-white wing-bars, double black breast-band, olive upperparts, white supercilium and whitish underparts. Migrant winter straggler to riverine forest in extreme southern Sind. Normally inhabits tropical evergreen rainforest.

2 WHITE WAGTAIL *Motacilla alba* 19 cm
a MALE BREEDING, **b** MALE NON-BREEDING and **c** JUVENILE *M. a. alboides*, **d** MALE BREEDING and **e** 1ST-WINTER *M. a. personata*, and **f** MALE *M. a. dukhunensis* Extremely variable. Head pattern, and grey or black mantle, indicate racial identification of breeding males. Non-breeding and first-winter birds often not racially distinguishable. Never has head pattern of White-browed. *M. a. dukhunensis* is the commonest subspecies; a very widespread winter visitor throughout the Indus plains and less common throughout Baluchistan. *M. a. personata* breeds on gravelly streams throughout northern areas and in the Kurram Valley, wintering in northern Punjab and NWFP foothills. *M. a. alboides* breeds sparingly from northern Chitral to Baltistan, with some wintering in the northen plains of Punjab and NWFP. Intermediates between *alboides* and *personata* sometimes overlap on summer breeding grounds and interbreed.

3 WHITE-BROWED WAGTAIL *Motacilla maderaspatensis* 21 cm
a ADULT and **b** JUVENILE Large black-and-white wagtail. Head black with white supercilium, and black mantle. Juvenile has brownish-grey head, mantle and breast, with white supercilium. Sings from February to October, repeated at intervals, *kwrita-kwrita-tse-e-e-tsi-ee-tu-uh*. Sedentary resident, occurring sparingly along major rivers of NWFP and Punjab, but more frequent in northern Punjab along major canals and on Salt Range lakes. Absent from Baluchistan and Sind. Avoids mountainous areas.

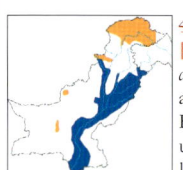

4 CITRINE WAGTAIL *Motacilla citreola* 19 cm
a MALE BREEDING, **b** FEMALE, **c** JUVENILE and **d e** 1ST-WINTER *M. c. calcarata*, and **f** MALE BREEDING *M. c. citreola* Broad white wing-bars in all plumages. Male breeding has yellow head and underparts, and black or grey mantle. Female breeding and adult non-breeding have broad yellow supercilium continuing around ear-coverts, grey upperparts, and mainly yellow underparts. Juvenile lacks yellow, and has brownish upperparts, buffish supercilium (with dark upper edge) and ear-covert surround, and spotted black gorget. First-winter has grey upperparts; distinguished from Yellow by white surround to ear-coverts, dark border to supercilium, pale brown forehead, pale lores, all-dark bill, and white undertail-coverts; by early November, has yellowish supercilium, ear-covert surround and throat. *M. c. calcarata* is a common partial resident and winter migrant, breeding in northern inner Himalayan valleys alongside lakes and streams up to the alpine zone (3900 m) and, sparingly, in wet areas in Baluchistan, with some migrating to central Asia; winters throughout the Indus plains, near rivers or lakes, typically foraging over open water on emergent vegetation. *M. c. citreola* is more plentiful in Sind, only as a winter visitor, and is more adaptable to forage in irrigated green crops and small seepage zones.

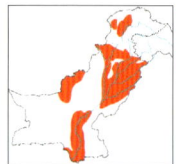

5 YELLOW WAGTAIL *Motacilla flava* 18 cm
a MALE BREEDING, **b** FEMALE, **c** JUVENILE and **d** 1ST-WINTER *M. f. beema*, **e** MALE *M. f. leucocephala*, **f** MALE *M. f. melanogrisea*, and **g** MALE and **h** 1ST-WINTER *M. f. thunbergi* Four races occur in Pakistan. Male breeding has olive-green upperparts and yellow underparts, with considerable variation in coloration of head depending on race. Female is extremely variable, but often has some features of breeding male. First-winter birds typically have brownish-olive upperparts, and whitish underparts with variable yellowish wash; in some races they can closely resemble Citrine, but have narrower white supercilium that does not continue around ear-coverts. (See also Table 7 on p.246.) Common double passage migrant, mostly wintering in India, with a few remaining in Sind. Often in flocks and always close to wetlands or irrigated green crops. Four subspecies conspicuous on spring passage when in breeding plumage. *M. f. beema* and *melanogrisea* commonest, *M. f. thunbergi* slightly less so, and *M. f. leucocephala* rare. Encountered from late August, with early April return passage. *M. f. thunbergi* mostly in Baluchistan and NWFP, *M. f. beema* mostly in Punjab, and *M. f. melanogrisea* throughout Indus plains.

1 GREY WAGTAIL *Motacilla cinerea* 19 cm

a MALE BREEDING, **b** FEMALE and **c** JUVENILE Longer-tailed than other wagtails. In all plumages, shows white supercilium, grey upperparts, and yellow vent and undertail-coverts. Male has black throat when breeding. Frequent to locally common breeding bird, in the vicinity of mountain streams of western Baluchistan north to Chitral and east to Baltistan. The majority winter in well-forested tracts in India. For nesting, avoids larger, broader mountain rivers favoured by White Wagtail, or marshy grassy areas favoured by Citrine Wagtail. A few overwinter in relict patches of riverine forest near seasonal lakes along the Indus basin and major river valleys of Gilgit and Baltistan.

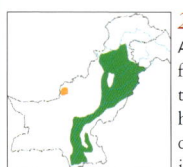

2 PADDYFIELD PIPIT *Anthus rufulus* 15 cm

ADULT Well-streaked upperparts and breast; lores usually look pale. When flushed, has comparatively weak, rather fluttering flight. Flight display is a fluttering ascent with repeated *tissip-tissip* notes. Parachute descent with wings held high accompanied by accelerando *tit-tit-tit* whistles. Has a *chip-chip-chip* flight call. Common resident throughout the Indus plains, some moving northwards into foothill areas to breed. Avoids dry desert, favouring grassy margins of lakes, seepage zones around barrage headworks and areas with extensive leguminous fodder crops.

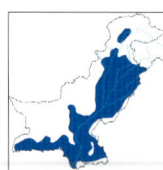

3 TAWNY PIPIT *Anthus campestris* 16 cm

a ADULT, **b** 1ST-WINTER and **c** JUVENILE Adult and first-winter have plain or faintly streaked upperparts and breast; lores are dark. Juvenile is more heavily streaked. Overall coloration is paler, more sandy grey than that of Paddyfield. Loud *tchilip* or *chep* call is quite different from Paddyfield's. Common winter visitor, preferring uncultivated barren tracts, from the foothills of NWFP, Sind and Punjab, including eastern desert regions. Largely absent from Baluchistan, except across Mekran.

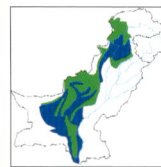

4 LONG-BILLED PIPIT *Anthus similis* 20 cm

ADULT Considerably larger than Tawny, with very large bill and shorter-looking legs. Dark lores. Greyish upperparts and warm buff colour to unstreaked or only lightly streaked underparts. Undulating circling display flight is accompanied by spaced *chirrit-chirrit-teeweeh-pr-chirrrit-teeweh* etc. Call is a sparrow-like *chirp*. Common resident, breeding in dry rocky areas throughout Baluchistan, NWFP, lower Chitral and across Hazara District to the Murree Hills up to 2400 m. Avoids forest, preferring bare open hillsides in the north and low stony slopes in Sind Kohistan.

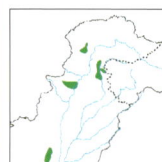

5 UPLAND PIPIT *Anthus sylvanus* 17 cm

a b ADULT Large, heavily streaked pipit with short and broad bill, and rather narrow, pointed tail feathers. Fine black streaking on underparts, whitish supercilium; ground colour of underparts varies from warm buff to rather cold and grey. Sings often from a tree perch, a thin, high-pitched 2-noted repeated whistle *whit-tree-whit-see*. Call is a sparrow-like *chirp*. Scarce, largely sedentary species, occurring in widely separated small populations from the Suleiman Hills of northeast Baluchistan, Kohat District, Swat, Hazara District and the Murree Hills. Prefers warmer southern unforested hill slopes (1300–2500 m).

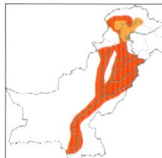

1 TREE PIPIT *Anthus trivialis* — 15 cm

a ADULT *A. t. trivialis* and **b** ADULT *A. t. haringtoni* Buffish-brown to greyish ground colour to upperparts (lacking olive cast of Rosy and Olive-backed pipits), and buffish (rather than olive) fringes to greater coverts, tertials and secondaries. Call is an abrupt *teez*. (See Appendix for differences from Olive-backed.) Display is a parachute descent accompanied by rapid warbling, terminating in a drawn-out *tsee-tsee-tsee*. Locally frequent summer breeder in northern areas, preferring damper alpine meadows just above the tree-line. Marked spring and autumn passage throughout the Indus basin, avoiding dry mountainous regions or extensive desert. Winters in India. Generally breeds at lower altitudes then Rosy Pipit in same areas.

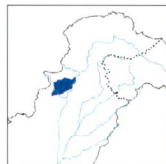

2 MEADOW PIPIT *Anthus pratensis* — 15 cm

ADULT Subtle differences from Tree are slimmer build, slimmer and weaker bill, and less boldly streaked breast but more boldly streaked flanks. Lacks prominent white supercilium, broad white wing-bars and distinct greenish edges to tertials and secondaries of Rosy Pipit. Call is a soft *sip-sip-sip*, most similar to that of Rosy Pipit. Rare winter visitor to extreme northwest NWFP, favouring grassy lake margins or fields.

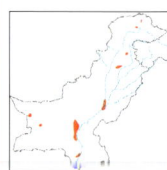

3 RED-THROATED PIPIT *Anthus cervinus* — 15 cm

a MALE BREEDING and **b** 1ST-YEAR Adult has reddish throat and upper breast, which tend to be paler on female and on autumn/winter birds. First-year has heavily streaked upperparts, pale mantle 'braces', well-defined white wing-bars, strongly contrasting blackish centres and whitish fringes to tertials, pronounced dark malar patch, and boldly streaked breast and (especially) flanks. Call is a drawn-out *seeee*. Rare double passage migrant, occurring mainly in northern Punjab, especially the Salt Range plateau and northern areas. Forages on grassy lake margins or open swampy areas.

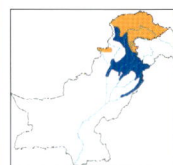

4 ROSY PIPIT *Anthus roseatus* — 15 cm

a b ADULT BREEDING and **c** ADULT NON-BREEDING Always has boldly streaked upperparts, olive cast to mantle, and olive to olive-green edges to greater coverts, secondaries and tertials. Adult breeding has mauve-pink wash to underparts. Non-breeding plumage has heavily streaked underparts and dark lores. Display includes much singing from prominent perches, interspersed with typical song *tit-tit-tit-teedle-teedle* during ascent flight, followed by *sweet-sweet-sweet* during parachute descent. Call is a weak *seep-seep*. Abundant resident, breeding throughout northern areas above the tree-line (up to 4500 m). Winters in outer foothills and valleys wherever there is damp grassland.

5 WATER PIPIT *Anthus spinoletta* — 15 cm

a ADULT BREEDING and **b** ADULT NON-BREEDING In all plumages, has lightly streaked upperparts, lacks olive-green on wing, has dark legs and usually has pale lores; underparts less heavily marked than on Rosy and Buff-bellied pipits. Orange-buff wash to supercilium and underparts in breeding plumage. Highly gregarious and common winter visitor, preferring damp grassy areas, rice stubbles, lake margins and seepage areas alongside major irrigation canals. Occurs from sea coast to far northern valleys and northwestern Baluchistan.

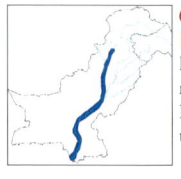

6 BUFF-BELLIED PIPIT *Anthus rubescens* — 15 cm

a ADULT BREEDING and **b** ADULT NON-BREEDING In all plumages, has lightly streaked upperparts, lacks olive-green on wing, and has pale lores; underparts more heavily streaked, upperparts darker and legs paler compared with Water Pipit. Winter visitor to marshes and damp cultivation from Kohat south through the Indus plains to Karachi.

ACCENTORS AND MOUNTAIN FINCHES, PLATE 88

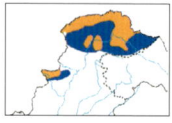

1 **ALPINE ACCENTOR** *Prunella collaris* 15.5–17 cm
ADULT Black barring on throat, grey breast and belly, and black band across greater coverts. Scarce resident, breeding at high altitudes in inner mountains beyond monsoon influence (3650–5200 m). Winters in adjacent valleys down to 1300 m.

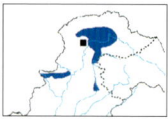

2 **ALTAI ACCENTOR** *Prunella himalayana* 15–15.5 cm
ADULT White throat, with black gorget and spotting in malar region; white underparts, with rufous mottling on breast and flanks. Uncommon winter visitor in small flocks to northern foothill areas. Occasionally breeds in alpine pastures above 3700 m.

3 **ROBIN ACCENTOR** *Prunella rubeculoides* 16–17 cm
ADULT Uniform grey head, rusty-orange band across breast, and whitish belly. Frequent resident, breeding in bleak alpine bush-dotted valleys near streams, particularly in Little Deosai, Baltistan and Astor, Gilgit (3600–4500 m). Winters down to 2000 m.

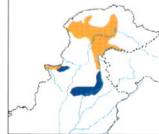

4 **RUFOUS-BREASTED ACCENTOR** *Prunella strophiata* 15 cm
a ADULT and **b** 1ST-WINTER? Rufous band across breast, white-and-rufous supercilium, blackish ear-coverts, and streaking on neck sides and underparts. Locally common resident, breeding in the subalpine scrub zone and preferring grassy moister slopes (2700–3600 m). Descends in winter to outer foothill valleys and the eastern escarpment of the Punjab Salt Range down to 600 m.

5 **BROWN ACCENTOR** *Prunella fulvescens* 14.5–15 cm
ADULT White supercilium, faintly streaked upperparts, and pale orange-buff underparts. (See Appendix for differences from Radde's Accentor.) Locally common resident in drier inner mountain ranges from Chitral to Baltistan, inhabiting bush-dotted rocky slopes (3200–4800 m). Winters in adjacent main valleys down to 1220 m.

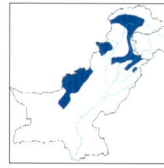

6 **BLACK-THROATED ACCENTOR**
Prunella atrogularis 14.5–15 cm
a ADULT and **b** 1ST-WINTER FEMALE Orange-buff supercilium and submoustachial stripe, and black throat. Some have indistinct (or lack) black on throat, which instead is whitish; note heavily streaked mantle. (See Appendix for differences from Radde's Accentor.) Locally common winter visitor throughout northwestern Baluchistan, Kohat, Chitral and Gilgit main valleys, south to the Punjab Salt Range. Winters in semiarid hilly areas with stunted juniper and acacia up to 2600 m.

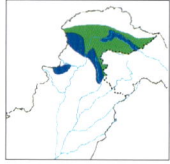

7 **PLAIN MOUNTAIN FINCH** *Leucosticte nemoricola* 15 cm
a ADULT BREEDING and **b** JUVENILE Told from Brandt's by boldly streaked mantle with pale 'braces', and distinct patterning on wing-coverts (dark-centred, with well-defined wing-bars). Juvenile is a warmer rufous buff than adult. Locally abundant resident, breeding on moister alpine slopes from Chitral to Baltistan (3200–4500 m). Highly gregarious, with winter flocks numbering up to 2000. Descends to outer foothills down to 1500 m.

8 **BRANDT'S MOUNTAIN FINCH**
Leucosticte brandti 16.5–19 cm
a MALE BREEDING and **b** 1ST-SUMMER Unstreaked to lightly streaked mantle, and rather uniform wing-coverts. More striking white panel on wing and more prominent white edges to tail compared with Plain. Adult breeding has sooty-black head and nape, and brownish-grey mantle with poorly defined streaking. Male has pink on rump. Uncommon but gregarious resident in higher, bleaker alpine valleys, north of the mountains frequented by Plain (3900–5200 m). Winters in main valleys of northern Hazara and Chitral to Baltistan down to 3200 m.

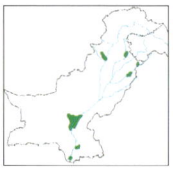

1 BLACK-BREASTED WEAVER *Ploceus benghalensis* 14 cm
a MALE BREEDING and **b** ADULT NON-BREEDING Breeding male has yellow crown and black breast-band. In other plumages, breast-band less distinct or restricted to small patches at sides, and may show indistinct streaking on lower breast and flanks; head pattern as on female/non-breeding Streaked, except crown, nape and ear-coverts are more uniform; rump is also indistinctly streaked and contrasts with streaked mantle. Frequent resident, occurring in scattered colonies for breeding. Always in reed-beds near lakes, rivers or canals. Less common in Sind than in NWFP and Punjab.

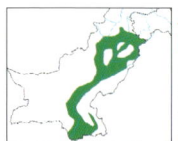

2 STREAKED WEAVER *Ploceus manyar* 14 cm
a MALE BREEDING and **b** ADULT NON-BREEDING Breeding male has yellow crown, dark brown head, and heavily streaked underparts. Other plumages typically show boldly streaked underparts; can be only lightly streaked on underparts, when best told from Baya by combination of yellow supercilium and neck-patch, heavily streaked crown and dark malar and moustachial stripes. Abundant resident on the Indus floodplain, breeding in reed-beds and most numerous in rice-growing tracts.

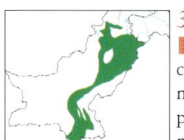

3 BAYA WEAVER *Ploceus philippinus* 15 cm
a MALE BREEDING and **b** ADULT NON-BREEDING Breeding male has yellow crown, dark brown ear-coverts and throat, unstreaked yellow breast, and yellow on mantle and scapulars. Female/non-breeding birds usually have unstreaked buff to pale yellowish underparts; can show streaking as prominent as on some poorly marked Streaked, but generally has less distinct and buffish supercilium, lacks yellow neck-patch, and lacks pronounced dark moustachial and malar stripes. Head pattern of some (non-breeding males?) can, however, be rather similar to Streaked. Locally abundant in the Indus basin, though not so closely tied to reed-beds as Streaked. Like all weavers, disperses widely outside the breeding season.

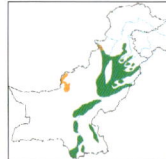

4 RED AVADAVAT *Amandava amandava* 10 cm
a MALE BREEDING, **b** FEMALE and **c** JUVENILE Breeding male is mainly red with white spotting. Non-breeding male and female have red bill, red rump, and white tips to wing-coverts and tertials. Juvenile lacks red in plumage; has buff wing-bars, pink bill-base, and pink legs and feet. Frequent resident in scattered colonies adjacent to wetlands, rivers and canals in Sind, Punjab, NWFP and Baluchistan. Has suffered severe decline due to cagebird trade.

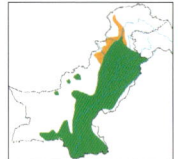

5 INDIAN SILVERBILL *Lonchura malabarica* 11–11.5 cm
a ADULT and **b** JUVENILE Adult has white rump and uppertail-coverts, black tail with elongated central feathers, and rufous-buff barring on flanks. Widespread common resident throughout lowland areas, including mountain valleys. Prefers uncultivated land with patches of thorn scrub and cane grass.

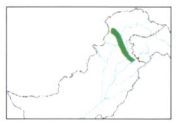

6 SCALY-BREASTED MUNIA *Lonchura punctulata* 10.7–12 cm
a ADULT and **b** JUVENILE Adult has chestnut throat and upper breast, and whitish underparts with dark scaling. Juvenile has brown upperparts and buffish underparts; bill black. Frequent resident, confined to a narrow belt along the Himalayan foothills and adjacent plains from Swat south to Lahore.

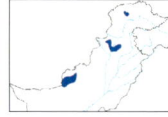

7 CHAFFINCH *Fringilla coelebs* 16 cm
a MALE NON-BREEDING and **b** FEMALE White wing-bars; lacks white rump. Male has blue-grey crown and nape, orange-pink face and underparts, and maroon-brown mantle. Female is dull, with greyish-brown mantle and greyish-buff underparts. Irregular, infrequent winter visitor to NWFP and the Punjab Salt Range. Commoner in northern Baluchistan, where birds with a distinctly bluish mantle occur.

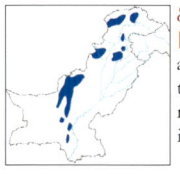

8 BRAMBLING *Fringilla montifringilla* 16 cm
a MALE BREEDING, **b** MALE NON-BREEDING and **c** FEMALE White rump and belly, and orange scapulars, breast and flanks. Patterning of head and mantle vary with sex and feather wear. Locally common winter visitor throughout mountainous regions of western Pakistan, most numerous in Baluchistan. Often in flocks feeding in snow.

FINCHES, PLATE 90

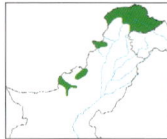

1 **FIRE-FRONTED SERIN** *Serinus pusillus*   12.5 cm
a MALE, **b** FEMALE and **c** JUVENILE Adult has blackish head with scarlet forehead. Juvenile has cinnamon-brown head. Abundant resident in higher mountain ranges of northern Baluchistan, the upper Kurram Valley and throughout northern mountains. Frequents upper limit of tree-line and both moist alpine and drier rocky slopes (2400–4700 m). Moves to adjacent valleys in winter (1600 m).

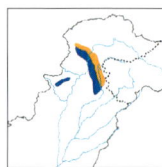

2 **YELLOW-BREASTED GREENFINCH**
Carduelis spinoides   14 cm
a MALE, **b** FEMALE and **c** JUVENILE Yellow supercilium and underparts, dark ear-coverts and malar stripe, and yellow patches on wing. Juvenile is heavily streaked. Locally frequent resident from Swat, southwest to the Neelum Valley. Prefers edges of forest, terraced cultivation and warmer southern-facing slopes (1500–3300 m). Erratic movement to outer hills in winter (1000 m).

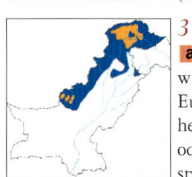

3 **EURASIAN GOLDFINCH** *Carduelis carduelis*   13–15.5 cm
a MALE and **b** JUVENILE Red face (lacking on juvenile), and black-and-yellow wings with white tertial markings. Lacks the black crown and cheek-stripe of the European population (and treated by some authorities as separate species, Grey-headed Goldfinch *Carduelis caniceps*). Locally common resident but erratic in its occurrence, breeding in the inner drier Himalayas from Chitral to Baltistan, also sparsely in higher mountain ranges in Baluchistan. Prefers thinly scattered trees on edge of forests (2100–3600 m). Winters down to adjacent plains, with numbers augmented by central Asian winter visitors.

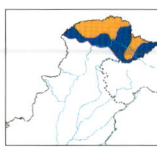

4 **TWITE** *Carduelis flavirostris*   13–13.5 cm
a MALE and **b** FEMALE Rather plain and heavily streaked, with small yellowish bill, buff wing-bars, and white edges to remiges and rectrices. Male has pinkish rump. Locally frequent resident, confined to the highest northern plateaux regions and valleys, nesting in open alpine habitats (2000–4700 m). Winters in adjacent lower valleys down to 1400 m.

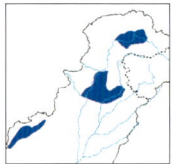

5 **EURASIAN LINNET** *Carduelis cannabina*   13–14 cm
a MALE BREEDING, **b** MALE NON-BREEDING and **c** FEMALE Told from Twite by greyish crown and nape that contrast with a browner mantle, whitish lower rump and uppertail-coverts, and larger greyish bill. Forehead and breast are crimson on breeding male. Frequent but irregular winter visitor in flocks to north-western Baluchistan, southern Gilgit, and Peshawar and Mardan districts; also the Punjab Salt Range. Favours open stony areas with few trees or bare stubble.

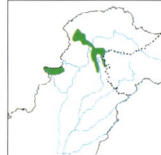

6 **SPECTACLED FINCH** *Callacanthis burtoni*   17–18 cm
a MALE and **b** FEMALE Black wings with white tips to feathers. Blackish head with red (male) or orange-yellow (female) 'spectacles'. Juvenile has browner head with buff eye-patch and buffish wing-bars. Scarce resident in Safed Koh, southwest Chitral and across northern Hazara District to the Neelum Valley. Breeds in the upper limit of coniferous forest in the Himalayas (2400–3350 m). Limited altitudinal movement in winter.

7 **CRIMSON-WINGED FINCH**
Rhodopechys sanguinea   15–18 cm
a MALE and **b** FEMALE Stout yellowish bill, pink on wing, dark cap, streaked brown mantle, and brown breast and white belly. Rare nomadic visitor to extreme northwestern borders of Chitral. Frequents bare stony slopes and alpine grassland.

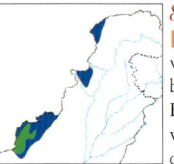

8 **DESERT FINCH** *Rhodospiza obsoleta*   14.5–15 cm
a MALE BREEDING and **b** FEMALE Pink edges to wing-coverts and secondaries, white edges to remiges and outer tail feathers, and black centres to tertials. Adult breeding has black bill. Male has black lores. Frequent resident in northwestern Baluchistan, with irregular movements to drier hills and valleys in NWFP and western Chitral (1500–2400 m). Described as very common in the early 20th century, but has since suffered persecution for the aviary trade. Despite its common name, it prefers habitats near trees or water.

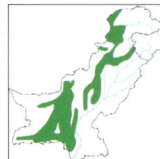

1 **TRUMPETER FINCH** *Bucanetes githagineus*   14–15 cm
a MALE and **b** JUVENILE Stocky, with very stout pinkish or yellow bill. Comparatively uniform wings and tail show traces of pink (except on juvenile). Locally common, highly nomadic resident from the sea coast to the far north. Inhabits dry rocky, hilly country, avoiding sand-dune desert and high-altitude plateau country favoured by Mongolian Finch. Singing male does indeed sound like a little toy trumpet.

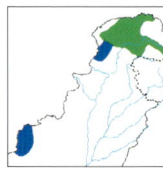

2 **MONGOLIAN FINCH** *Bucanetes mongolicus*   14–15 cm
a MALE and **b** FEMALE Whitish panels at bases of greater coverts and secondaries (which can be rather indistinct), whitish outer edges to tail, and less stout bill than Trumpeter's. Locally frequent nomadic resident in dry hilly areas and upland plateaux in the far north, less common in Baluchistan. Often sympatric with Trumpeter Finch.

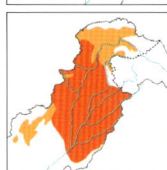

3 **COMMON ROSEFINCH** *Carpodacus erythrinus*   14.5–15 cm
a MALE and **b** FEMALE *C. e. roseatus*, **c** MALE and **d** FEMALE *C. e. erythrinus*
Male has red head, breast and rump. Female has streaked upperparts and underparts, and double wing-bar. Migrant *erythrinus* has less red in male, and female is less heavily streaked, compared with resident *roseatus*. (See also Table 8 on p.247.) Locally common resident, and abundant spring migrant through Punjab and NWFP. Breeds in higher ranges in Baluchistan and Safed Koh, and from Chitral to Baltistan (up to 3900 m). Prefers terraced cultivation or slopes with good vegetative cover.

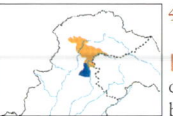

4 **PINK-BROWED ROSEFINCH**
Carpodacus rodochrous   14–15 cm
a MALE and **b** FEMALE Male has deep pink supercilium and underparts, and crimson crown. Female has buff supercilium and fulvous ground colour to rump, belly and flanks. (See also Table 8 on p.247.) Locally migratory scarce resident, breeding at the upper limit of the tree-line in monsoon-influenced Himalayas, Indus Kohistan, south to the lower Neelum Valley. Small flocks winter in the Margallas.

5 **WHITE-BROWED ROSEFINCH** *Carpodacus thura*   17 cm
a MALE and **b** FEMALE Large size. Male has pink-and-white supercilium, pink rump and underparts, and heavily streaked brown upperparts. Female has prominent supercilium with dark eye-stripe, ginger-brown throat and breast, and olive-yellow rump. (See also Table 8 on p.247.) Scarce resident, breeding in Safed Koh, northern Hazara District, and from Chitral to Baltistan. Typically on precipitous alpine slopes above 3000 m. Descends only in winter, to lower slopes down to 2400 m.

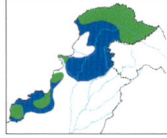

6 **RED-MANTLED ROSEFINCH**
Carpodacus rhodochlamys   18 cm
a MALE and **b** FEMALE Large size and large bill. Male has pink supercilium and underparts, and pale grey-brown upperparts with pinkish wash. Female is pale grey and heavily streaked, with indistinct supercilium. Frequent resident from Baluchistan and NWFP to the far north. Prefers drier mountain regions than White-browed and Pink-browed, but sympatric with Common Rosefinch in juniper forests of Baluchistan (2900–3500 m). (See also Table 8 on p.247.) Winters down to foothills and the Punjab Salt Range.

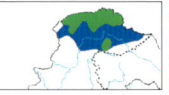

7 **GREAT ROSEFINCH** *Carpodacus rubicilla*   19–20 cm
a MALE and **b** FEMALE Large size and long tail. Male has pale pink head and underparts, and pale sandy-grey and lightly streaked upperparts. Female lacks supercilium, and has lightly streaked sandy-brown upperparts and underparts. (See also Table 8 on p.247.) Common resident; high-altitude arid rocky areas from Chitral to Baltistan, up to 5000 m. Winters in adjacent valleys to 1500 m, and attracted to Sea Buckthorn.

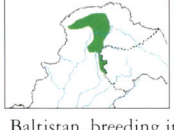

8 **RED-FRONTED ROSEFINCH** *Carpodacus puniceus*   20 cm
a MALE and **b** FEMALE Large size, conical bill and short tail. On male, red of plumage contrasts with brown crown, eye-stripe and upperparts. Female lacks supercilium, is heavily streaked, and may show yellow/olive rump. (See also Table 8 on p.247.) Scarce sedentary resident from northern Swat through Gilgit and Baltistan, breeding in high alpine valleys, 3300–4400 m. Remains year-round at high altitudes, preferring better vegetated slopes than Great Rosefinch.

FINCHES AND BUNTINGS, PLATE 92

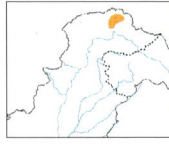

1 **RED CROSSBILL** *Loxia curvirostra* 16–17 cm
a MALE, **b** FEMALE and **c** JUVENILE Dark bill with crossed mandibles. Male is rusty red. Female is olive-green, with brighter rump. Juvenile is heavily streaked; mandibles are not initially crossed. Irregular irruptions occur westwards across the Himalayas into Pakistan according to seed-setting on Pine cones. Usually found in small parties; recorded only from Gilgit, feeding on Blue Pine and spruce cones.

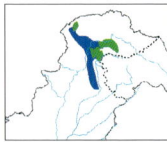

2 **ORANGE BULLFINCH** *Pyrrhula aurantiaca* 14 cm
a MALE, **b** FEMALE and **c** JUVENILE Male has orange head and body and orange-buff wing-bars. Female has grey crown and nape. Juvenile is similar to female, but lacks grey on head. Scarce resident, confined to Himalayan forest between 2700 m and 3200 m. Rare in Chitral and Swat, more plentiful in northern Hazara District, across to the Neelum Valley. Winter visitor only to Murree Hills.

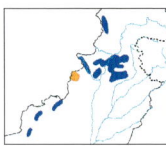

3 **HAWFINCH** *Coccothraustes coccothraustes* 16–18 cm
MALE Stocky, short-tailed and huge-billed. Mainly orange-brown, with pale wing-covert band and black chin. Locally frequent winter visitor, restricted to dry scrub forest foothill tracts. Occasional breeder in the Torghar Valley, Zhob District, Baluchistan (2300 m).

4 **BLACK-AND-YELLOW GROSBEAK**
Mycerobas icterioides 22 cm
a MALE and **b** FEMALE Male is black and yellow (lacking orange cast to mantle of similar Collared Grosbeak, which occurs in India). Female has pale grey head, mantle and breast, and peachy-orange rump and belly. Common resident in temperate Himalayan forest, 1800–3350 m. Safed Koh, and from Chitral south to the Murree Hills.

5 **SPOT-WINGED GROSBEAK** *Mycerobas melanozanthos* 22 cm
a MALE and **b** FEMALE Male has black mantle and rump, and white markings on wings. Female is yellow, boldly streaked with black; wing pattern is similar to male's. Scarce and erratic visitor to Murree Hills and Neelum Valley. Mainly present in late summer, feeding on Himalayan Bird Cherry. Occasionally breeds about 2700 m in Silver Fir forest.

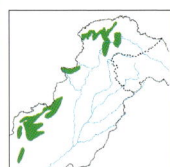

6 **WHITE-WINGED GROSBEAK** *Mycerobas carnipes* 22 cm
a MALE and **b** FEMALE White patch in wing. Male is dull black and olive-yellow, with yellowish tips to greater coverts and tertials. Female is similar, but black of plumage is replaced by sooty grey. Locally frequent sedentary resident, highly dependent on juniper berries. Forested mountains of northern Baluchistan, and subalpine zone in Safed Koh and northern areas where creeping juniper is plentiful (2400–4000 m).

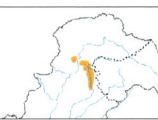

7 **CRESTED BUNTING** *Melophus lathami* 17 cm
a MALE, **b** FEMALE and **c** 1ST-WINTER MALE Always has crest and chestnut on wing and tail; tail lacks white. Frequent summer breeding visitor to the outer lower Himalayan foothills. Typically found in subtropical Pine zone and open terraced hillsides, 1000–1800 m. Winters mainly eastwards in Siwaliks, India.

8 **PINE BUNTING** *Emberiza leucocephalos* 17 cm
a MALE BREEDING, **b** MALE NON-BREEDING and **c** FEMALE Chestnut rump and long tail. Male has chestnut supercilium and throat, and whitish crown and ear-covert spot; pattern obscured in winter. Female has greyish supercilium and nape/neck sides, dark border to ear-coverts, usually some rufous on breast/flanks, and white belly. Frequent winter visitor to main valleys in northern mountain areas, from Chitral to Baltistan. Less numerous but widespread in Baluchistan (typically around 1500 m). Occasionally strays to Peshawar and Rawalpindi Districts.

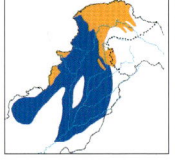

9 **ROCK BUNTING** *Emberiza cia* 16 cm
a MALE and **b** FEMALE Male has grey head, black crown-sides and border to ear-coverts, and rufous underparts. Female is duller. Common resident, breeding from northern NWFP and Chitral to Baltistan, and south through Hazara District into the Murree Hills. Usually inhabits the upper limit of coniferous forest or subalpine and alpine zones (2100–4200 m). Winters widely in Baluchistan and Punjab, not extending into Sind.

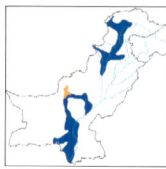

1 GREY-NECKED BUNTING *Emberiza buchanani* — 15 cm

a MALE, **b** FEMALE and **c** 1ST-WINTER Pinkish-orange bill, plain head and whitish eye-ring. Adult has blue-grey head, buffish submoustachial stripe and throat, and rusty-pink underparts. First-winter and juvenile often have only slight greyish cast to head and buffish underparts; light streaking on breast. (See Appendix for differences from Ortolan Bunting.) A scarce wintering bird throughout dry rocky hilly country in Sind, Punjab and Baluchistan, with majority of population double passage migrants. Breeds irregularly in dispersed colonies in central Baluchistan in *Artemisia* steppe.

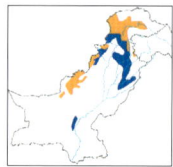

2 WHITE-CAPPED BUNTING *Emberiza stewarti* — 15 cm

a MALE BREEDING, **b** MALE NON-BREEDING and **c** FEMALE Male has grey head, black supercilium and throat, and chestnut breast-band; pattern is obscured in winter. Female has rather plain head with pale supercilium; crown and mantle uniformly and diffusely streaked, and underparts finely streaked and washed with buff. Common resident, breeding on lower warmer slopes throughout northern areas (1200–3000 m). Occurs in Baluchistan and Safed Koh in juniper or Chilghoza forest up to 2600 m. Winters on adjacent plains in dry scrub forest.

3 HOUSE BUNTING *Emberiza striolata* — 13–14 cm

a MALE and **b** FEMALE Has black eye-stripe and moustachial stripe, and white supercilium and submoustachial stripe; throat and breast streaked, and underparts brownish buff with variable rufous tinge. Female is duller than male, with less striking head pattern. Lacks prominent white on tail (Rock has much white on outer tail feathers) and has orange lower mandible (bill all grey on Rock). Frequent resident throughout barren hilly desert areas west of the Indus, from the sea coast to NWFP, with limited nomadism in search of fallen seeds.

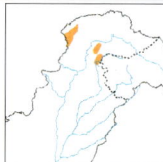

4 CHESTNUT-EARED BUNTING *Emberiza fucata* — 16 cm

a MALE, **b** FEMALE and **c** 1ST-WINTER Adult has chestnut ear-coverts, black breast streaking, and chestnut on breast sides. Some birds are nondescript; plain head with warm brown ear-coverts and distinctive pale eye-ring. Rare summer breeding bird in 3 disjunct populations: Chitral, Kaghan Valley and Neelum Valley, in open terraced cultivated slopes 1800–2400 m. Winters in adjacent foothills.

5 BLACK-HEADED BUNTING
Emberiza melanocephala — 16–18 cm

a MALE BREEDING, **b** MALE NON-BREEDING, **c** WORN FEMALE and **d** IMMATURE Male has black on head and chestnut on mantle. Female when worn may show ghost pattern of male; fresh female almost identical to Red-headed, but indicative features include rufous fringes to mantle and/or back, slight contrast between throat and greyish ear-coverts, and more uniform yellowish underparts. Immature has buff underparts and yellow undertail-coverts. Common double passage migrant in flocks, in lower Sind and Baluchistan; migrates in a northwesterly direction.

6 RED-HEADED BUNTING *Emberiza bruniceps* — 16 cm

a MALE BREEDING, **b** MALE NON-BREEDING, **c** FRESH FEMALE and **d** IMMATURE Smaller than Black-headed, with shorter, more conical bill. Male has rufous on head and yellowish-green mantle. Female when worn may show rufous on head and breast, and yellowish to crown and mantle, and is distinguishable from female Black-headed. Fresh female has paler throat than breast, with suggestion of buffish breast-band; forehead and crown are often virtually unstreaked (indicative features separating it from Black-headed). Immature is often not separable from Black-headed but may exhibit some of the features mentioned above. Abundant double passage migrant in flocks, in northern Baluchistan, the whole of Punjab and NWFP. Occasionally breeds in central Baluchistan. Prefers dry treeless steppe country.

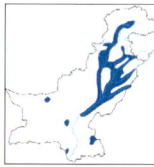

7 REED BUNTING *Emberiza schoeniclus* — 14–15 cm

a MALE BREEDING and **b** FEMALE Male has black head and white submoustachial stripe; obscured by fringes when fresh. Female has buff supercilium, brown ear-coverts and dark moustachial stripe. Widespread and locally frequent winter visitor, occurring in small numbers wherever extensive patches of *Phragmites* or *Saccharum* reed/grasses occur, around major lakes and riverine tracts. Less frequent in Baluchistan and southern Sind.

Appendix: Vagrants and Extirpated species

Species that have been recorded only once or on a few occasions, and extirpated species, are described below.

Fulvous Whistling-duck *Dendrocygna bicolor* 51 cm
Larger than Lesser, with bigger, squarer head and larger bill. Adult told from adult Lesser by warmer rufous-orange head and neck, dark blackish line down hind-neck, dark striations on neck, more prominent streaking on flanks, indistinct chestnut-brown patch on forewing, and white band across uppertail-coverts. Freshwater wetlands. A scattering of records, but only one in recent decades.

Mute Swan *Cygnus olor* 125–155 cm
Adult is white and has orange bill with black base and knob. Juvenile is mottled sooty brown, and has grey bill with black base. Large rivers and lakes. A few authenticated records at the beginning of the 20th century.

Whooper Swan *Cygnus cygnus* 140–165 cm
Adult is white with black and yellow bill; has yellow of bill extending as wedge towards tip. Juvenile is smoky grey, with pinkish bill. Longer neck and more angular head shape than Tundra. Lakes and large rivers. Most recent records from Zangi Nawar Lake, Baluchistan.

Tundra Swan *Cygnus columbianus* 115–140 cm
Adult is white with black and yellow bill; yellow of bill is typically an oval-shaped patch. Juvenile is smoky grey, with pinkish bill. Smaller in size, and with shorter neck and more rounded head, compared with Whooper. Lakes and large rivers. Two records from the early part of the 20th century.

Lesser White-fronted Goose *Anser erythropus* 53–66 cm
Smaller and more compact, with squarer head and stouter bill, compared with Greater White-fronted (like that species, has orange-pink bill, and orange legs and feet). White frontal band of adult extends onto forehead. Both adult and juvenile have yellow eye-ring, and darker head and neck than Greater. Wet grassland and lakes. Occasional records up until 1965. Globally threatened (Vulnerable).

Falcated Duck *Anas falcata* 48–54 cm
Male has bottle-green head with maned hind-neck, and elongated black-and-grey tertials; shows pale grey forewing in flight. Female has rather plain greyish head, a dark bill, dark spotting and scalloping on brown underparts, and greyish-white fringes to exposed tertials; shows greyish forewing and white greater-covert bar in flight, but does not show striking white belly (compare with female Eurasian Wigeon). Lakes and large rivers. About five records up until the 1920s.

Baikal Teal *Anas formosa* 39–43 cm
Grey forewing and broad white trailing edge to wing in flight in both sexes (recalling Northern Pintail). Male has striking head pattern of black, yellow, bottle-green and white; also black-spotted pinkish breast, black undertail-coverts and chestnut-edged scapulars. Female superficially resembles female Common Teal; has dark-bordered white loral spot and buff supercilium that is broken above eye by dark crown; some females have white half-crescent on cheeks. Lakes and large rivers. One record from Sind in 1878 and one from Punjab in the mid-20th century. Globally threatened (Vulnerable).

Baer's Pochard *Aythya baeri* 41–46 cm
Resembles Ferruginous Pochard, but has greenish cast to dark head and neck, which contrast with chestnut-brown breast. White patch on fore-flanks visible above water. Female and immature male have duller head and breast than adult male. Female has dark iris, and pale and diffuse chestnut-brown loral spot. Large rivers and lakes. One record from Punjab in January 1957. Globally threatened (Vulnerable).

Greater Scaup *Aythya marila* 40–51 cm
Larger and stockier than Tufted Duck, and lacking any sign of crest. Male has grey upperparts that contrast with black rear-end. Immature male is duller. Female has broad white face-patch, which is less extensive on juvenile/immature. Female usually has greyish-white vermiculations ('frosting') on upperparts and flanks. Large lakes and rivers. Occasionally recorded in Pakistan.

Long-tailed Duck *Clangula hyemalis* 36–47 cm
Small, stocky duck with stubby bill and pointed tail. Swims low in water and partly opens wings before diving. Both sexes show dark upperwing and underwing in flight. Winter male is mainly white; has dark cheek-patch and breast, and long tail. Female and immature male are variable;

usually have dark crown, and pale face with dark cheek-patch. Lakes and large rivers. Occasionally recorded in first half of 20th century.

Yellow-rumped Honeyguide *Indicator xanthonotus* 15 cm
Golden-yellow forehead, back and rump, streaked underparts, and square, blackish tail. Inner edges of tertials are white, and form parallel lines down back. Male has pronounced yellow malar stripes. Near Giant Rock Bee nests on cliffs, and adjacent forest. Formerly recorded from the Murree Hills but not since 1908 and now presumed extirpated in Pakistan.

Lesser Yellownape *Picus chlorolophus* 27 cm
Medium-sized, mainly green woodpecker with tufted yellow nape, scarlet and white markings on head, and barring on underparts. Broadleaved forest and secondary growth. A single record in 1982 in the Murree Hills.

Oriental Pied Hornbill *Anthracoceros albirostris* 55–60 cm
Medium-sized black-and-white hornbill. Head and neck black. Tail mainly black with white tips to outer feathers. Both sexes have cylindrical casque. Female has smaller casque than male, and has black at tip of bill. Calls include a variety of loud, shrill, nasal squeals and raucous chucks. Mature broadleaved forest with fruiting trees. One record, photographed, 13 March 1990, in Mirpur, Poonch.

Black-capped Kingfisher *Halcyon pileata* 30 cm
Large, strikingly patterned kingfisher with black cap, white collar, purplish-blue upperparts and pale orange underparts. Shows white wing-patch in flight. Call is a ringing cackle, *kikikikikiki*. One record, photographed, 18 October 1998, on the Hab river on the Sind Kohistan border.

Large Hawk Cuckoo *Hierococcyx sparverioides* 38 cm
Larger than Common Hawk Cuckoo, with browner upperparts, strongly barred underparts, and broader tail-banding. Juvenile has barred flanks and broad tail-banding; head dark grey on older birds. Call is a shrill *pee-pee-ah...pee-pee-ah*, which is repeated, rising in pitch and momentum, climaxing in hysterical crescendo. Broadleaved forest. Recorded from the Murree Hills in first half of 20th century.

Snowy Owl *Nyctea scandiaca* 53–66 cm
Very large, mainly white owl, with variable dark markings. Open country. Recorded twice, in 1876 and 1989.

Boreal Owl *Aegolius funereus* 24–26 cm
A smallish owl, a little larger than Little Owl. Greyish in coloration, with heavy whitish markings on upperparts. Underparts diffusely barred with grey. Angular head shape, with greyish facial discs bordered with black. Call is a soft *po-po-po*. A single record of a calling male in subalpine juniper scrub and stunted trees, in February 1986, in the Barpu Glacier valley on the Nagar Hunza border.

Orange-breasted Green Pigeon *Treron bicincta* 29 cm
Male has orange breast, bordered above by lilac band, grey hind-crown and nape, and green mantle. Female has yellow cast to breast and belly, and grey hind-crown and nape. Central tail feathers of both sexes are grey. Forest. Two recorded in Karachi in 1938.

Great Bustard *Otis tarda* 75–105 cm
Very large, stocky bustard. In all plumages, has greyish head and upper neck, cinnamon lower neck, cinnamon upperparts with bold black barring, and white underparts. Patterning of white on wing differs from that of other bustards (secondaries are black, and white is most prominent on the greater coverts). Male is larger than female, with thicker neck and with more extensive white on wing (tertials and wing-coverts show more white). Grassland and crops. Occasionally recorded, mainly in the Peshawar area in the first half of the 20th century. Globally threatened (Vulnerable).

Corn Crake *Crex crex* 27–30 cm
A stocky crake with stout pinkish bill and legs. Has rufous chestnut on wings (especially prominent in flight), greyish fore-neck and breast, and rufous-brown and white barring on flanks. Juvenile has buffish, rather than grey, neck and breast. Grassland and crops. Only one record, from Gilgit in 1881.

Red Knot *Calidris canutus* 23–25 cm
Stocky, with short, straight bill, and rather short legs. Smaller than Great Knot, and with straighter bill. Plumages are similar to those of Curlew Sandpiper, although has barred rump and uppertail-coverts. Adult breeding is brick-red on underparts, with upperparts patterned with rufous and black. Adult non-breeding is whitish on underparts and uniform grey on upperparts; stocky shape, short bill and barred rump are best features separating it from other calidrids. Juvenile has buff fringes and dark subterminal crescents to upperparts. The only record is a specimen purportedly collected by Meinertzhagen in Baluchistan in 1920; many significant Meinertzhagen records from the Indian subcontinent have recently been discredited and this record should be treated with caution.

Sharp-tailed Sandpiper *Calidris acuminata* 17–21 cm
Recalls Wood Sandpiper in shape. Has rufous crown (indistinct in winter) and prominent white supercilium. Adult non-breeding is greyish above, with breast-band of fine streaking. Adult breeding has dark markings over entire underparts, with arrow-head markings on flanks, and bright rufous fringes to feathers of mantle and scapulars. Juvenile is similar to adult breeding but has buff wash to lightly streaked breast. Freshwater wetlands. Two records from Pakistan, the last in 1970.

Red Phalarope *Phalaropus fulicaria* 20–22 cm
Typically seen swimming. Stockier than Red-necked Phalarope, with stouter bill that is often pale or yellowish at base. Adult breeding has red neck and underparts, and white face-patch. Adult non-breeding has more uniform and paler grey mantle, scapulars and rump than Red-necked. Juvenile has dark upperparts that are evenly fringed with buff (lacking mantle and scapular stripes of Red-necked). A single inland record, 18–21 August 1987, from Rawal Lake, Islamabad.

European Golden Plover *Pluvialis apricaria* 26–29 cm
Very similar in plumage to Pacific Golden Plover, but stockier, with shorter and stouter bill and shorter legs. Underwing-coverts and axillaries largely white. At rest, primaries do not extend beyond tail as in Pacific. In flight, toes do not project beyond tail (noticeable projection on Pacific). Breeding plumage is similar to that of Pacific, although shows more white on sides of breast and flanks. In non-breeding plumage, supercilium is usually less distinct and is rather plain-faced, compared with Pacific. Grassland and mud on lake shores and in estuaries. Three records prior to 1919.

Eurasian Dotterel *Charadrius morinellus* 21 cm
A stocky plover, with comparatively short, yellow legs. Very distinctive in breeding plumage, with prominent white supercilium, narrow white breast-band, and chestnut and blackish belly. In juvenile and non-breeding plumage, belly is buffish, but retains distinctive supercilium and suggestion of narrow breast-band. Dry open country. A single record, the only one from the Indian subcontinent, photographed in February 1990 in the Hab Valley, Sind Kohistan.

Oriental Pratincole *Glareola maldivarum* 23–24 cm
Adult breeding has black-bordered creamy-yellow throat and peachy-orange wash to underparts (patterning much reduced in non-breeding plumage). Shows red underwing-coverts in flight. Lacks white trailing edge to wing, which is the best feature separating it from Collared Pratincole. Adult has less pronounced fork to tail than in Collared, with tail-tip falling short of closed wings at rest. Dry bare ground near wetlands. Only reliably recorded in East Nara, when found nesting with Collared Pratincoles at the end of the 19th century.

Red-headed Vulture *Sarcogyps calvus* 85 cm
Adult is blackish with bare reddish head and cere, white patches at base of neck and upper thighs, and reddish legs and feet; in flight, greyish-white bases to secondaries show as broad panel. Juvenile is browner, with white down on head. Pinkish coloration to head and feet, white patch on upper thighs, and whitish undertail-coverts are best features separating it from other vultures. Open country and well-wooded hills. Previously occurred regularly in the Punjab and Sind, but now thought to be only a rare straggler with a record from Hyderabad District, Sind in 2007. Globally threatened (Critical).

Crested Serpent Eagle *Spilornis cheela* 56–74 cm
Broad, rounded wings. Soars with wings held forward and in a pronounced V. Adult has broad white bands across wings and tail; hooded appearance at rest, with yellow cere and lores, and white spotting on brown underparts. Juvenile has blackish ear-coverts, yellow cere and lores, whitish head and underparts, narrower barring on tail (than adult), and largely white underwing with fine dark barring and dark trailing edge. When soaring, frequently utters loud whistling cry. Forest and well-wooded country. A few records from Pakistan, mainly in the Murree Hills and Rawalpindi area, but none since that of a bird in captivity in 1975.

Pied Harrier *Circus melanoleucos* 41–46.5 cm
Male has black head, upperparts and breast, white underbody and forewing, and black median-covert bar. Female has white uppertail-covert patch, dark-barred greyish remiges and rectrices, pale leading edge to wing, pale underwing and whitish belly. Juvenile has pale markings on head, rufous-brown underbody, white uppertail-covert patch, and dark underwing with pale patch on primaries. Open grassland and cultivation. Rare visitor, with one recent record in winter, Punjab Salt Range lakes.

Besra *Accipiter virgatus* 29–36 cm
Small. Upperparts darker than those of Shikra, and prominent gular stripe and streaked breast should separate it from Eurasian Sparrowhawk; underwing strongly barred compared with that of Shikra. Male has dark slate-grey upperparts, broad blackish gular stripe, and bold rufous streaking

on breast and barring on belly. Female browner on upperparts, with blackish crown and nape. Juvenile told from juvenile Shikra by darker, richer brown upperparts, broader gular stripe and broader tail-barring. A specimen is reported to have been collected in the Murree Hills.

Indian Spotted Eagle *Aquila hastata* 60–65 cm

Now recognised, once again, as a distinct species from Lesser Spotted Eagle *Aquila pomarina* (which is now best considered extralimital). As Greater Spotted, Indian Spotted is a stocky, medium-sized eagle with rather short and broad wings, a buzzard-like head with a comparatively fine bill, and a rather short tail. The wings are angled down at carpals when gliding and soaring. Adult is similar in overall appearance to Greater Spotted, and field characters are poorly understood. Has a wider gape than Greater Spotted, with thick 'lips', and with gape-line extending well behind eye (reaching to below centre of eye in Spotted). A possible additional feature used to distinguish adults in the field is the paler brown lesser underwing-coverts, which contrast with rest of underwing (Greater Spotted typically has uniform dark underwing-coverts). Juvenile is more distinct from juvenile Greater Spotted; spotting on upperwing-coverts is less prominent, tertials are pale brown with diffuse white tips (dark with bold white tips in Greater Spotted), uppertail-coverts are pale brown with white barring (white in Greater Spotted), and underparts are paler light yellowish brown with dark streaking. In some plumages it can resemble Steppe Eagle – differences mentioned between Greater Spotted are likely to be helpful for separation (although gape-line is also long in Steppe). Three records, one from the Murree Hills, where a nest was claimed and female collected (presumably this species) in March 1988 at Rawal Lake, and one at Dera Ismail Khan in 1996. Globally threatened (Vulnerable).

Red-necked Grebe *Podiceps grisegena* 40–50 cm

Slightly smaller than Great Crested, with stouter neck, squarer head and stockier body, which is often puffed up at rear-end. Black-tipped yellow bill. Black crown extends to eye, and has dusky cheeks and fore-neck in non-breeding plumage. White cheeks and reddish fore-neck in breeding plumage. Lakes. Recorded on three occasions, the most recent in 1987.

Horned Grebe *Podiceps auritus* 31–38 cm

Bill is stouter and does not appear upturned as it does in Black-necked. Head shape is different: triangular-shaped, with crown peaking at rear. Has two white patches on upperwing, with white patch on wing-coverts (usually lacking on Black-necked). White cheeks contrast with black crown and white fore-neck in non-breeding plumage (cheeks and fore-neck dusky in Black-necked). Yellow ear-tufts and rufous neck and breast in breeding plumage. Lakes. Recorded on four occasions, the most recent in 1986.

Pygmy Cormorant *Phalacrocorax pygmeus* 45–55 cm

On average is larger and bulkier than Little Cormorant. Adult breeding has chestnut head and upper neck (becoming nearly black prior to breeding), with more profuse white plumes than breeding Little. Non-breeding and immature are very similar to Little, although tend to be browner on body and have more extensive whitish mottling on breast and belly; in adult non-breeding, chin and throat are whitish, gradually merging into brown of fore-neck (on Little, chin is more clearly demarcated). Inland waters. A single record, from southern Baluchistan, in 1909.

Asian Openbill *Anastomus oscitans* 68 cm

Stout, dull-coloured 'open bill'. Largely white (breeding) or greyish white (non-breeding), with black flight feathers and tail; legs usually dull pink, brighter in breeding condition. Juvenile has brownish-grey head, neck and breast, and brownish mantle and scapulars that are slightly paler than the blackish flight feathers. At a distance in flight, best told from White Stork by dull-coloured bill and black tail. Marshes and lakes. Recorded as breeding up until the 1930s, but only occasional records since then.

Woolly-necked Stork *Ciconia episcopus* 75–92 cm

Stocky, largely blackish stork with 'woolly' white neck, black 'skullcap', and white vent and undertail-coverts. In flight, upperwing and underwing appear entirely dark. Juvenile is similar to adult but has duller brown body and wings, and feathered forehead. Flooded fields, marshes and lakes. Breeding records from the 20th century, with known pairs up until 1980, but possibly now extirpated in Pakistan.

Greater Adjutant *Leptoptilos dubius* 120–150 cm

Larger than similar Lesser Adjutant (although this species has not been recorded in Pakistan), with stouter, conical bill that has a convex ridge to the culmen. Adult breeding has bluish-grey mantle, silvery-grey panel across greater coverts, greyish or brownish underwing-coverts, grey undertail-coverts, and more extensive white neck-ruff. Further, it has a blackish face and forehead (with appearance of dried blood), and a neck pouch (visible only when inflated). Adult non-breeding and immature have darker grey mantle and inner wing-coverts, and brown greater coverts (which barely contrast with rest of wing); immature has brownish (rather than pale) iris. Marshes. Globally threatened (Vulnerable). A scattering of records, the last being a captive bird caught in the late 1950s.

Red-throated Loon *Gavia stellata* 53–69 cm
Upturned-looking bill and rounded head. In non-breeding plumage, is paler grey and whiter than similar Black-throated (although this species has not been recorded in Pakistan). Grey of crown and hind-neck is paler and less extensive compared with Black-throated and does not contrast so strongly with white of ear-coverts and fore-neck. Red throat and uniform grey-brown upperparts in breeding plumage. A single record from inshore waters of the Makran coast.

Short-tailed Shearwater *Puffinus tenuirostris* 41–43 cm
Sooty brown, with pale grey underwing-coverts. Much as the very similar Sooty Shearwater *P. griseus* (not recorded in Pakistan), but has shorter bill and steeper forehead, shorter extension of rear body and tail behind wings, and pale panel on underwing tends to be less striking. A single record in 1899 from the Makran coast.

Long-tailed Broadbill *Psarisomus dalhousiae* 28 cm
Long tail, which is often held cocked. Green, with black cap, and yellow 'ear' spot and throat. Juvenile has green cap. Has a loud, piercing call, *pieu-wieuw-wieuw-wieuw*..., usually 5 to 8 notes on the same pitch. Moist broadleaved forest. A single record, photographed on 21 August 1990, near Kotli, Poonch.

Woodchat Shrike *Lanius senator* 17 cm
Adult is striking, with chestnut hind-crown and nape, black upperparts with white scapular and rump, and white underparts. The sexes are similar although the female is duller. Shows prominent white patch in primaries in flight. Juvenile is heavily scaled; told from juvenile Red-backed Shrike by paler, more prominently scaled crown and nape, whitish centres to scapulars, pale rump, and whitish patch at base of primaries. Scrub and open country. The only substantiated record is from near Karachi in 2002.

Great Grey Shrike *Lanius excubitor* 25 cm
Lacks black forehead and has pale grey mantle. Very similar to *pallidirostris* race of Southern Grey Shrike (which is paler, with pale lores and pale bill, and lacks prominent black forehead compared with the more widely distributed *lahtora* race of Southern Grey), but has broad white band at base of secondaries, and whitish rump and uppertail-coverts. Open country. Vagrants recorded from Baluchistan.

White-bellied Minivet *Pericrocotus erythropygius* 15 cm
A distinctive minivet with white wing-patch and orange on rump. Male has black head and upperparts, and white underparts with orange breast. Female has brown upperparts and white underparts. Dry open scrub and forest. Recorded only once, at Karachi in 1919.

Black-naped Monarch *Hypothymis azurea* 16 cm
Male mainly blue, flycatcher-like bird, with black nape and gorget. Female lacks these features and is duller, with grey-brown mantle and wings. Middle storey of broadleaved forest. Three records from Karachi, the most recent during the winter of 1979/80.

Bohemian Waxwing *Bombycilla garrulus* 18 cm
Mainly fawn-brown in coloration, and with prominent crest. Has black throat, and waxy red and yellow markings on wings and tail. Has starling-like appearance in flight, when often utters distinctive soft ringing trill. Open country with fruiting trees and bushes. Occasional winter occurrence, with the most recent, a small group, recorded in Karachi during a cold spell in 1969.

Redwing *Turdus iliacus* 22 cm
Smallish thrush with spotted/streaked underparts. Best told from Song Thrush by prominent whitish supercilium and red patch on flanks and underwing-coverts. One record, part of a dead bird, was picked up in Dehra Ismail Khan District, NWFP, in February 1989, and is now deposited with the Bombay Natural History Society.

Song Thrush *Turdus philomelos* 23 cm
Small size (much smaller than Mistle Thrush), olive-brown upperparts, and orange-buff wash to spotted breast. Head rather plain, lacking supercilium. Deciduous woodland. A single record in February 1984.

Eurasian Robin *Erithacus rubecula* 13 cm
Rotund and well-proportioned robin, with olive-brown upperparts, including tail, and rufous-orange throat and breast. Two recent records. Photographed on 13 February 2000 in Islamabad, Margalla Hills, and seen on 28 December 2000 in Dunga Gali, Murree Hills.

Common Nightingale *Luscinia megarhynchos* 16 cm
Much as Bluethroat in shape and behaviour, but with longer tail. Rather nondescript, with greyish olive-brown upperparts and greyish-white underparts, and whitish throat. Has rufous uppertail-coverts and long rufous tail, indistinct head markings, and pale fringes to wing-coverts and remiges. Bushes, usually feeding on the ground. Around 6 records, mainly from Baluchistan.

Purple-backed Starling *Sturnus sturninus* 19 cm
Male has pale grey head and underparts, purplish-black hind-crown patch and mantle, and white tips to median coverts and rear scapulars. Female and juvenile are duller; wing-bars and tips to scapulars are less prominent in juvenile. Open wooded areas. A single record, in July, from Chitral in the early 20th century.

Azure Tit *Parus cyanus* 13.5 cm
Mainly whitish head, with dark stripe through eye and band across nape, grey mantle, white underparts, and broad white wing-bar and tips to tertials. Similar in appearance to Yellow-breasted Tit, but lacking yellow on breast of that species. Seasonally dry river-bed bushes. A single record from Hunza.

Yellow-breasted Tit *Parus flavipectus* 13 cm
Similar in appearance to Azure Tit, but has yellow breast and greyish head; chin and centre of throat greyish (show as faint bib). Dense thickets. A single record from Chitral.

Eurasian Reed Warbler *Acrocephalus scirpaceus* 13 cm
Very similar in appearance to Blyth's Reed Warbler, but with longer primary projection. In worn (breeding) plumage, it is rather pale brown above and white below (Blyth's Reed is more grey-toned), whilst in fresh (non-breeding) plumage it is more rufescent (Blyth's Reed is more olive-toned). Supercilium is less prominent, bill is longer, and primary projection is longer, compared with Paddyfield and Blunt-winged warblers. Song is a varied series of scratchy notes, and less musical (lacking clear, rising notes of Blyth's Reed). Reed-beds. Unconfirmed record of breeding in Malezai Lora, Chagai District, in 1942.

Great Reed Warbler *Acrocephalus arundinaceus* 19 cm
Differs from Clamorous Reed in having a shorter, stouter bill, longer primary projection, and shorter-looking tail. Primary projection is roughly equal to length of tertials, with 8 or 9 exposed primary tips visible beyond the tertials (primary projection in Clamorous is two-thirds of tertial length, with 6 or 7 exposed tips visible beyond the tertials). Reed-beds. A single record from Baluchistan in 1911.

Dusky Warbler *Phylloscopus fuscatus* 11 cm
Brown upperparts and whitish underparts, with buff flanks. Prominent supercilium, and hard *chack* call. Call and rather skulking habits are good features separating it from Common Chiffchaff. Bushes and long grass, especially near water. See also Table 2 on p.242. A single record from Islamabad in 1996.

Blyth's Leaf Warbler *Phylloscopus reguloides* 11 cm
Very similar to Western Crowned Warbler, although smaller and more compact. It is also brighter above, head pattern is more clearly defined with yellowish supercilium, and is yellower on underparts. Wing-bars appear broader and more prominent, with dark panel across greater coverts. Broadleaved and coniferous forests. See also Table 5 on p.244. Recorded as breeding in the Murree Hills in the early 20th century, and then one further record from there in 1984.

Bristled Grassbird *Chaetornis striatus* 20 cm
Streaked upperparts and fine streaking on lower throat. Has shorter, stouter bill and less prominent supercilium than Striated Grassbird, and a shorter and broader tail with buffish-white tips. Short grassland with scattered bushes and some tall vegetation. Three records: a breeding pair near Lahore in 1914, one in Sind in 1976, and two singing by the Kabul River in 1999. Globally threatened (Vulnerable).

Chestnut-crowned Laughingthrush *Garrulax erythrocephalus* 28 cm
An olive-brown laughingthrush with a chestnut crown, dark scaling/spotting on the mantle and breast, olive-yellow wings and olive-yellow tail sides. Undergrowth in broadleaved forest. Recorded in the Murree Hills in the late 19th century but presumed now to be extirpated in Pakistan.

Bearded Parrotbill *Panurus biarmicus* 16–17 cm
A round-bodied, long-tailed babbler-like bird with a small yellow bill. Has white edges to primaries, which form a wing-panel. Male has grey head with black moustache. Female and juvenile have plain buff head. Has distinctive, ringing *ping-ping* call. Reed-beds. Recorded once on the Indus river in January 1927.

Barred Warbler *Sylvia nisoria* 15 cm
Large *Sylvia* with stout bill, and pale edges and tips to tertials and wing-coverts in all plumages.

Adult has greyish upperparts, yellow iris and variable dark barring on whitish underparts. First-winter has greyish olive-brown upperparts and buffish underparts, with barring on undertail-coverts and, occasionally, on flanks. Bushes. Two records, the most recent from Baltistan in July 1933.

Richard's Pipit *Anthus richardi*   17 cm
Very similar in appearance to Paddyfield Pipit, with well-streaked upperparts and breast. Best told from that species by larger size, and loud *schreep* call. Pale lores help separate it from juvenile Tawny Pipit. When flushed, typically gains height with deep undulations (compared with Paddyfield). Status in Pakistan unclear due to similarity with Paddyfield Pipit.

Olive-backed Pipit *Anthus hodgsoni*   15 cm
Distinguished from Tree Pipit by greenish-olive cast to upperparts (although becoming more grey with wear), and greenish-olive fringes to greater coverts, tertials and secondaries. Typically, has more striking head pattern than Tree. *A. h. yunnanensis*, which is a winter visitor to the subcontinent, is much less heavily streaked on upperparts than resident nominate race and Tree. Open forest and shrubberies. A flock of about 15 was recorded and photographed in woodland northeast of Islamabad suburbs, 29 January–4 February 1996.

Radde's Accentor *Prunella ocularis*   15–16 cm
Very similar in appearance to Brown Accentor, but has dark brown crown (paler than ear-coverts in Brown), heavily streaked mantle, spotted malar stripe, buffish breast-band and streaked flanks. Some show no dark markings on malar and flanks. Confusion is also possible with first-winter Black-throated Accentor, which may not show dark throat, but has white supercilium and darker crown and ear-coverts. Shrubs in mountainous areas. A single sighting, plus a probable record from the Quetta area, Baluchistan.

Ortolan Bunting *Emberiza hortulana*   16 cm
Similar in appearance to Grey-necked Bunting, with pinkish-orange bill, plain head and prominent eye-ring. Adult has olive-grey head and breast, and yellow submoustachial stripe and throat. Female is similar, with streaking on crown and breast. First-winter and juvenile are more heavily streaked on mantle, malar region and breast than Grey-necked; submoustachial stripe and throat are buffish, but often with touch of yellow, which helps separate them from Grey-necked. Orchards and open woodland. Records from Gilgit in the 19th century and one from Chaman, Baluchistan (based on an old specimen in the British Museum of Natural History).

Yellow-breasted Bunting *Emberiza aureola*   15 cm
A strongly marked bunting with yellow coloration to underparts in all plumages. Male has black face and throat and chestnut breast-band, with white patch on wing-coverts. In fresh plumage, head pattern is masked by pale fringes to feathers and is similar to that of female; chestnut breast-band remains as a distinctive feature. Female has strikingly patterned head, with prominent supercilium and dark border to ear-coverts, and has boldly striped mantle; shows white median-covert bar. Juvenile is similar to female but has streaked underparts. See comparison with Chestnut Bunting. Cultivation and grassland. Recorded only on the Makran coast in November 1901.

Chestnut Bunting *Emberiza rutila*   14 cm
Small size and yellow on underparts in all plumages. Male has chestnut head and breast, which may be obscured due to pale fringes in first-winter plumage. Female and first-winter have yellowish underparts, and are most likely to be confused with female/first-winter Yellow-breasted Bunting. Chestnut is smaller and slighter, with finer bill; also has less striking head pattern (with indistinct buffish supercilium, and lacking prominent dark border to ear-coverts), more finely streaked mantle, bright and unstreaked chestnut rump, and little or no white on tail. Told from female Black-faced Bunting (not recorded in Pakistan) by buff throat and brighter yellow underparts (lacks bold dark flank streaking of that species), chestnut rump, and little or no white on tail. Bushes in cultivation, forest clearings. Recorded only from Chitral in April 1902.

Corn Bunting *Miliaria calandra*   18 cm
A large, stocky bunting with a stout bill and comparatively short tail. Heavily streaked brownish upperparts and buffish underparts. Cultivation. A single record from the Punjab of a small flock in November 1917.

TABLES

Table 1. Nightjars

Species	Size/structure	General coloration	Crown/nape	Scapulars/coverts	Primaries	Tail	White throat-patch
Grey Nightjar I = *C. i. indicus* (south of Himalayan range) H = *C. i. hazarae* (Himalayan range)	Medium. Well proportioned, longish wings and tail, and large head	I – Uniform cold grey to grey-brown, heavily marked with black. H – Dark grey-brown, heavily marked with black	Variable. Heavily to very heavily marked with black drop-shaped streaks. Some have irregular patches of rufous on nape. Streaking less regular and more extensive than on Large-tailed	Scapulars heavily but irregularly marked with black, usually lacking well-defined buff or white edges (although pronounced pale edges to these feathers may be prominent in some). Coverts have variable greyish-white to buffish spotting; usually poorly defined in I, more rufous in H	Male has small white spots on 3 or 4 primaries. On female, spots either lacking or small and rufous	I – Male variable, usually with white at tips of all but central tail feathers and diffuse greyish margin at end. H – Male has more distinct blackish margin to white tips. Female of both races lacks white in tail	Large central white spot in male, or buff in female
Eurasian Nightjar	Medium. Long wings and tail, and small head	Grey with neat lanceolate streaking	Regular, bold, black lanceolate streaking	Bold lanceolate streaking to scapulars, with buff outer edge. Well-defined, regular pale buff spots on coverts	Male has large white spots on 3 primaries and female has no spots	Outermost 2 tail feathers have broad white tips in male. No white tips in female	Indistinct, but generally complete white throat-crescent
Sykes's Nightjar	Small. Shortish wings and tail, and large head	Grey with buff mottling and restricted dark vermiculations. Some are as sandy as Egyptian	Variable, small dark arrow-head markings, more extensive than in Egyptian. Irregular buff spotting on nape gives suggestion of collar	Scapulars relatively unmarked with a few black inverted 'anchor-shaped' marks. Coverts with irregular and small buff markings	Male has large white spots on 3 or 4 primaries. Female primary spots are buffish	In male, 2 outermost pairs have broad white tips. In female, tail is unmarked or has buffish tip to outer tail feathers	Broken, large white patches on sides. Some have complete crescent
Indian Nightjar	Small. Short wings and tail, and small head	Grey, with bold buff, black and some rufous markings	Bold, broad, black streaking to crown. Nape marked with rufous buff, forming distinct collar	Bold triangular black centres and broad rufous-buff fringes to scapulars. Coverts with bold buff or rufous-buff spotting	Both sexes have small white or buffish spots on 4 primaries	Both sexes have broad white tips to outer 2 tail feathers	Generally broken. Large white patches on sides. Lacking in some
Large-tailed Nightjar	Large. Long-winged, and long, broad tail. Large head	More warmly coloured than Grey, with buff-brown tones, heavily marked with black and buff	Brownish grey with bold black streaks down centre. Diffuse pale rufous-brown band across nape	Scapulars have well-defined buff edges with bold wedge-shaped black centres. Coverts boldly tipped buff	Male has white spots on 4 primaries, which are lacking in female or replaced with smaller buff spots	Male has extensive white tips to 2 outermost feathers. Female has less extensive buff tips to outer 2 feathers	Large central white throat-patch in both sexes
Savanna Nightjar	Medium. Shortish wings and tail, and large head	Dark brownish grey, intricately patterned (without bold, dark streaking) but with variable rufous-buff markings	Variable. Some only finely vermiculated, others with black, 'arrow-head' markings and others with irregular-shaped black markings	Scapulars variably marked but most show rufous-buff outer web. Coverts variably marked with rufous-buff, showing as distinct spotting in some	Male has large white spots on 4 primaries, with buff to rufous-buff wash on female	Outer 2 tail feathers are mainly white in male, but not in female	Large white patches on sides

Table 2. Small to medium-sized Phylloscopus warblers, lacking wing-bars and crown-stripe
(+ = vagrant)

Species	Head pattern	Upperparts, including wings	Underparts	Call	Additional features
Common Chiffchaff	Whitish or buffish supercilium, and prominent crescent below eye	Greyish to brownish with olive-green cast to rump and edges of remiges and rectrices	Whitish with variable buffish or greyish cast to breast-sides and flanks	Plaintive *peu*, more disyllabic *sie-u*	Blackish bill and legs (compare with Greenish and Dusky warblers)
Mountain Chiffchaff	Whitish or buffish supercilium. Warm buff coloration to ear-coverts in fresh plumage	Brownish to greyish brown, lacking olive-green cast to rump. Edges of wing-coverts, remiges and rectrices are buffish. Bend of wing is usually whitish (usually brighter yellow in Common Chiffchaff)	Whitish, with warm buffish coloration to breast-sides and flanks in fresh plumage (usually more pronounced than in Common Chiffchaff)	Distinctly disyllabic *swe-eet* or *tiss-yip*; sometimes 3 syllables *tiss-yuitt*	Blackish bill and legs
Plain Leaf Warbler	Whitish supercilium	Greyish brown. Lacks greenish to upperparts; edges of remiges buffish	Whitish; buff on ear-coverts, breast-sides and flanks usually less apparent than in Mountain Chiffchaff	A hard *tak tak*, low-pitched *churr* or *chiip*, and a *twissa-twissa*	Very small; short-looking tail
Dusky Warbler +	Broad, buffish-white supercilium with strong, dark eye-stripe	Dark brown to paler greyish brown; never shows any greenish in plumage	White with buff on sides of breast and flanks	Hard *chack chack*	Pale brown legs, and orangish base to lower mandible. Typically skulking
Tickell's Leaf Warbler	Prominent yellow supercilium concolorous with throat, and well-defined eye-stripe	Dark greenish to greenish-brown upperparts, with greenish edges to remiges	Bright lemon-yellow underparts, lacking strong buff tones	A *chit*, or *sit*; not as hard as that of Dusky	
Sulphur-bellied Warbler	Prominent, bright sulphur-yellow supercilium, distinctly brighter than throat	Cold brown to brownish grey, lacking greenish tones, and with greyish edges to remiges	Yellowish buff with strong buff tones to breast and flanks, and sulphur-yellow belly	Soft *quip* or *dip*	Climbs about rocks, or nuthatch-like on tree trunks
Tytler's Leaf Warbler	Prominent fine white to yellowish-white supercilium, with broad, dark olive eye-stripe	Greenish, becoming greyer when worn	Whitish, with variable yellowish wash when fresh	A double *y-it*	Long, slender, mainly dark bill; shortish tail

Table 3. Medium-sized to large **Phylloscopus** *warblers, with narrow wing-bars, and lacking crown-stripe* Note that wing-bars may be missing when plumage is worn (when confusion is possible with species in Table 2)

Species	Head pattern	Upperparts, including wings	Underparts	Bill	Call	Other features
Greenish Warbler *P. t. viridanus*	Prominent yellowish-white supercilium, usually wide in front of eye and extends to forehead; eye-stripe usually falls short of base of bill	Olive-green, becoming duller and greyer when worn; generally lacks darker crown. Single narrow but well-defined white wing-bar	Whitish with faint yellowish suffusion	Lower mandible orangish, usually lacking prominent dark tip	Loud, slurred *chit-wee*	
Greenish Warbler *P. t. nitidus*	Prominent yellowish supercilium, and yellow wash to cheeks	Upperparts brighter and purer green than *viridanus*, with 1 or 2 slightly broader and yellower wing-bars	Strongly suffused with yellow, which can still be apparent in worn plumage	As *viridanus*	More trisyllabic than *viridanus*, a *chis-ru-weet*	
Large-billed Leaf Warbler	Striking yellowish-white supercilium that contrasts with broad, dark eye-stripe, and with greyish mottling on ear-coverts	Dark oily green with noticeably darker crown; 1 or 2 yellowish-white wing-bars	Dirty, often with diffuse streaking on breast and flanks and oily-yellow wash on breast and belly; can, however, appear whitish	Large and mainly dark, with orange at base of lower mandible; often has pronounced hooked tip	Loud, clear, whistled, upward-inflected *der-tee*	Large size

Table 4. Small **Phylloscopus** *warblers with broad wing-bars*

Species	Head pattern	Wing-bars	Rump and tail	Underparts	Call	Other features
Lemon-rumped Warbler	Yellowish crown-stripe that contrasts with dark olive sides of crown; yellowish-white supercilium	Double yellowish-white wing-bars	No white on tail; well-defined yellowish (sometimes almost whitish) rump	Uniform whitish or yellowish white	High-pitched *uist*	
Brooks's Leaf Warbler	Narrow, usually poorly defined yellowish crown-stripe; sides to crown barely darker than mantle; yellowish supercilium and wash to ear-coverts	Double yellowish wing-bars; bases of median and greater coverts do not form dark panel across wing as they do in Lemon-rumped	No white on tail. Ill-defined yellowish rump, often barely apparent	Yellowish throat; entire underparts washed with buffish yellow in fresh plumage	Monosyllabic, loud and piercing *chwee*, or *pseo* or *psee*	Brighter yellowish-olive upperparts compared with similar species. Yellowish-horn basal half of lower mandible (bill mainly dark in Hume's)
Hume's Warbler	Lacks well-defined crown-stripe, although can show diffuse paler line; broad buffish-white supercilium and cheeks	Broad, buffish or whitish greater-covert wing-bar; median-covert wing-bar tends to be poorly defined, but can be prominent	Lacks pale rump patch and does not have white in tail	White, often sullied with grey	A rolling, disyllabic *whit-hoo* or *visu visu*, and a flat *chwee*	Greyish-olive upperparts, with variable yellowish-green suffusion, and browner crown. Bill appears all dark and legs are normally blackish brown

*Table 5. Large **Phylloscopus** warblers with crown-stripe, prominent wing-bars, and large bill with orange lower mandible* (+ = vagrant)

Species	Head pattern	Upperparts, including wings	Underparts	Call	Additional features
Western Crowned Warbler	Greyish-white to pale yellow crown-stripe, contrasting with dusky-olive sides of crown, which may be darker towards nape; prominent dull yellow supercilium	Generally duller greyish green compared with Blyth's, with stronger grey cast to nape; wing-bars narrower and less prominent than Blyth's, because bases not so dark	Whitish, strongly suffused with grey, especially on throat and breast; can show traces of yellow on breast and belly	A repeated *chit-weei*	Larger and more elongated than Blyth's, with larger and longer bill
Blyth's Leaf Warbler +	Tends to be more striking than Western Crowned, with yellow supercilium and crown-stripe, contrasting with darker sides of crown	Usually darker and purer green than Western Crowned, although may be similar. Wing-bars are more prominent than Western and Eastern crowned, being broader and often divided by dark panel across greater coverts	Generally has distinct yellowish wash, especially on cheeks and breast	Constantly repeated *kee-kew-i*	

Table 6. Unstreaked Acrocephalus warblers and Hippolais warblers

Species	Bill/feet	Head pattern	Upperparts	Underparts	Additional features
Paddyfield	Shorter bill than that of Blyth's Reed, usually with well-defined dark tip to pale lower mandible. Yellowish-brown to pinkish-brown legs and feet	More prominent white supercilium, often broadening behind eye, becoming almost square-ended, with dark eye-stripe; supercilium can appear to be bordered above by diffuse dark line (supercilium less distinct on some)	More rufescent than Blyth's Reed. Typically shows dark centres and pale fringes to tertials. Greyer or sandier when worn, but usually retains rufous cast to rump	Warm buff flanks; underparts whiter when worn. Often shows whitish sides to neck	Typically looks longer-tailed than Blyth's Reed, with tail often held cocked
Blunt-winged	Longer and stouter bill than that of Paddyfield, with uniformly pale lower mandible or with dark shadow at tip	Shorter and less distinct supercilium than on Paddyfield, which typically barely extends beyond eye (occasionally on worn birds may extend as thin line behind eye); lacks dark border above supercilium and lacks prominent dark eye-stripe	As Paddyfield, but more olive-toned when fresh. Colder olive-brown when worn, but with more rufescent rump and uppertail-coverts. Shows more prominent tertial fringes than Blyth's Reed	Breast and flanks washed with buff when fresh	Longer tail, and shorter primary projection, compared with Paddyfield
Blyth's Reed	Bill longer than that of Paddyfield. Lower mandible either entirely pale or has diffuse dark tip	Comparatively indistinct supercilium; often does not extend beyond eye, or barely does so, and never reaches rear of ear-coverts. Lacks dark upper border to supercilium and dark eye-stripe	Tertials rather uniform. Generally colder olive-grey to olive-brown than Paddyfield. Noticeable warm olive cast to upperparts when fresh (more rufescent in first-winter)	Can have light buffish wash on flanks when fresh; otherwise cold whitish	Has shorter-looking, more rounded tail than Booted, and longer upper- and undertail-coverts; more skulking and lethargic than that species
Eurasian Reed	Bill longer than that of Paddyfield and Blunt-winged warblers	Very similar in appearance to Blyth's Reed; supercilium less prominent compared with those of Paddyfield and Blunt-winged warblers	In worn (breeding) plumage, is rather pale brown (Blyth's Reed is more grey-toned), whilst in fresh (non-breeding) plumage it is more rufescent (Blyth's Reed is more olive-toned)	White below in worn (breeding) plumage, and more buffish when fresh	Very similar in appearance to Blyth's Reed Warbler, but with longer primary projection (primary projection also longer compared with Paddyfield and Blunt-winged warblers)
Booted *H. c. rama*	Longer billed than that of *caligata*. Legs and feet paler and browner than those of Blyth's Reed	Supercilium more distinct than on Blyth's Reed and lores can appear pale	Paler and greyer than *caligata* and all Acrocephalus (although can be rather similar to Blyth's Reed)	Off-white	More arboreal than *caligata*; behaviour often *Phylloscopus*-like compared to Acrocephalus. Longer-looking square-ended tail than Acrocephalus, with shorter undertail-coverts
Booted *H. c. caligata*	Comparatively short and fine bill	Supercilium more prominent than on *rama*; can appear to have dark border	Warmer brown than *rama*. Fine whitish fringes to remiges and edges of outer tail feathers often apparent (also shown by *rama*)	Off-white	Rather *Phylloscopus*-like in appearance, often feeding on ground. Squarer tail and short undertail-coverts compared with Acrocephalus
Upcher's	Large and long bill	Supercilium short and barely extends beyond eye	Greyer than Acrocephalus and Booted, and flight feathers and tail often look noticeably darker. White edges and tips to outer rectrices, and prominent pale fringes to remiges and wing coverts	Whitish	Larger than Booted, with broader and fuller tail and longer primary projection. Has habit of flicking tail downwards

Table 7. Yellow Wagtails (breeding males only)

Subspecies	Head pattern
M. f. beema	Pale bluish-grey head, complete and distinct white supercilium, white chin, and usually a white submoustachial stripe that contrasts with yellow throat; ear-coverts are grey or brown, usually with some white feathers
M. f. leucocephala	Whole head to nape white, with a variable blue-grey cast on the ear-coverts and rear crown; chin is white, and throat is yellow, as are rest of underparts
M. f. melanogrisea	Black head, lacking any supercilium, and white chin and poorly defined submoustachial stripe that contrasts with yellow throat
M. f. thunbergi	Dark slate-grey crown with darker ear-coverts, lacking supercilium (although may show faint trace behind eye)

Table 8. Rosefinches

Female rosefinches

Species	Size/structure	Supercilium	Wing-coverts and tertials	Underparts	Upperparts	Other features
Common	Small and compact with stout, stubby bill	Lacking	Narrow whitish or buff tips to coverts, forming narrow double wing-bar	Whitish, with variable bold, dark streaking	Grey-brown with some dark streaking	Beady-eyed appearance
Pink-browed	Relatively small and compact	Prominent buff supercilium, contrasting with dark ear-coverts	Relatively uniform, lacking wing-barred effect	Heavily streaked, with strong fulvous wash from breast to undertail-coverts	Warm brownish buff with heavy, dark streaking	Fulvous coloration on rump
White-browed	Medium-large	Prominent long, white supercilium, contrasting, with dark lower border and rear of ear-coverts	Whitish to buff tips to median and greater coverts, forming narrow double wing-bar	White, heavily streaked, with racially variable ginger-buff wash to throat and breast	Mid- to dark brown with heavy dark streaking	Deep olive-yellow rump, heavily streaked
Red-mantled	Large, with stout, heavy bill and long tail	Rather faint whitish supercilium, weakly offset against greyish eye-stripe	Lacks wing-bars, but has indistinct paler tips to median and greater coverts	Whitish underparts, heavily streaked	Pale grey, heavily streaked	
Great	Large, with stout bill	Lacking	Relatively uniform pale grey-brown	Whitish, heavily streaked	Sandy brown with faint streaking	
Red-fronted	Very large, with rather long, conical bill	Lacking	Lacks wing-bars	Greyish, very heavily streaked, with variable pale yellow wash on breast	Very dark grey with bold, heavy, dark streaking	Rump and uppertail-coverts more olive than back

Male rosefinches

Species	Size/structure	General coloration	Supercilium	Wing-coverts and tertials	Underparts	Upperparts	Rump
Common	Small and compact, with stout, stubby bill	Bright geranium-red	Lacking	Rather indistinct bright red fringes to wing-coverts and tertials	Bright geranium-red, especially on throat and belly	Diffusely streaked bright geranium-red	Unstreaked, bright red
Pink-browed	Smallish and compact	Deep, warm pink	Prominent, deep pink, contrasting, with dark maroon-pink eye-stripe	Rather uniform, lacks prominent paler fringing to wing-coverts and tertials	Unstreaked, warm pink	Pinkish-brown mantle and back streaked darker; unstreaked or lightly streaked maroon-pink crown	Unstreaked, deep pink
White-browed	Medium-large	Brownish and pale pink	Prominent, long, pinkish with splashes of white, extends across forehead	Pronounced pale tips to median and greater coverts, forming narrow wing-bars	Pink, with some white streaking	Brown with conspicuous darker streaking	Pink, well-defined, contrasting with rest of upperparts
Red-mantled	Large, with stout, heavy bill and long tail	Pink and pale brownish grey	Prominent, pink	Very indistinct paler fringing	Pink, with some streaking	Brownish grey, with pink tinge, streaked darker	Uniform pink
Great	Large, with stout bill	Rose-pink and sandy grey	Lacking	Very indistinct paler fringing	Rose-pink, with large diffuse white spots	Sandy grey, washed pink, with narrow streaking	Uniform rose-pink
Red-fronted	Very large, with rather long, conical bill	Red and dark grey-brown	Short and red, extending across forehead	Very indistinct paler fringing	Red throat and breast, contrasting with dark-streaked grey-brown remainder of underside	Grey-brown, with bold, dark streaks	Deep pink, contrasting with rest of upperparts

INDEX

* = vagrant or extirpated

English Names

Accentor Alpine 222
  Altai 222
  Black-throated 222
  Brown 222
  Radde's** 240
  Robin 222
  Rufous-breasted 222
Adjutant Greater** 237
Avadavat Red 224
Avocet Pied 102

Babbler Black-chinned 204
  Common 206
  Green Shrike 206
  Jerdon's 204
  Jungle 206
  Large Grey 206
  Rusty-cheeked Scimitar 204
  Striated 206
  White-browed Shrike 206
  Yellow-eyed 204
Barbet Blue-throated 66
  Coppersmith 66
  Great 66
Bee-eater Blue-cheeked 68
  Blue-tailed 68
  European 68
  Green 68
Besra* 236
Bittern Black 144
  Cinnamon 144
  Great 144
  Little 144
  Yellow 144
Blackbird Eurasian 164
  Grey-winged 164
Bluethroat 172
Booby Masked 150
Brambling 224
Broadbill Long-tailed* 238
Bulbul Black 190
  Himalayan 190
  Red-vented 190
  White-eared 190
Bullfinch Orange 230
Bunting Black-headed 232
  Chestnut* 240

  Chestnut-eared 232
  Corn* 240
  Crested 230
  Grey-necked 232
  House 232
  Ortolan* 240
  Pine 230
  Red-headed 232
  Reed 232
  Rock 230
  White-capped 232
  Yellow-breasted* 240
Bushchat Grey 178
  Pied 178
  Stoliczka's 178
Bushlark Indian 208
  Singing 208
Bustard Great* 235
  Indian 86
  Little 86
  Macqueen's 86
Buttonquail Small 48
  Yellow-legged 48
Buzzard Common 130
  Long-legged 130
  White-eyed 130

Chaffinch 224
Chiffchaff Common 200
  Mountain 200
Chough Red-billed 156
  Yellow-billed 156
Chukar 48
Cisticola Zitting 194
Coot Common 90
Cormorant Great 140
  Indian 140
  Little 140
  Pygmy* 237
Coucal Greater 72
Courser Cream-coloured 108
  Indian 108
Crab-plover 102
Crake Baillon's 90
  Brown 90
  Corn* 235
  Little 90
  Ruddy-breasted 90
  Spotted 90
Crane Common 88

  Demoiselle 88
  Sarus 88
Crossbill Red 230
Crow Carrion 156
  House 156
  Large-billed 156
Cuckoo Common Hawk 70
  Eurasian 70
  Grey-bellied 72
  Indian 70
  Large Hawk* 235
  Lesser 70
  Oriental 70
  Pied 70
Cuckooshrike Black-winged 158
  Large 158
Curlew Eurasian 96

Darter 140
Dipper Brown 160
  White-throated 160
Dotterel Eurasian* 236
Dove Eurasian Collared 84
  European Turtle 84
  Laughing 84
  Oriental Turtle 84
  Red Collared 84
  Spotted 84
Drongo Ashy 160
  Black 160
Duck Comb 54
  Falcated* 234
  Long-tailed* 234
  Marbled 58
  Spot-billed 56
  Tufted 60
  White-headed 54
Dunlin 100

Eagle Black 124
  Bonelli's 134
  Booted 134
  Crested Serpent* 236
  Golden 132
  Greater Spotted 132
  Imperial 134
  Indian Spotted* 237
  Mountain Hawk 134
  Pallas's Fish 118

Short-toed Snake 124
Steppe 132
Tawny 132
White-tailed 118
Egret Cattle 142
Great 142
Intermediate 142
Little 142
Western Reef 142

Falcon Laggar 138
Peregrine 138
Red-necked 136
Saker 138
Sooty 136
Fantail White-browed 158
White-throated 158
Yellow-bellied 158
Finch Brandt's Mountain 222
Crimson-winged 226
Desert 226
Mongolian 228
Plain Mountain 222
Spectacled 226
Trumpeter 228
Flameback Black-rumped 64
Flamingo Greater 146
Lesser 146
Florican Lesser 86
Flowerpecker Thick-billed 212
Flycatcher Asian Brown 168
Blue-throated 170
Dark-sided 168
Grey-headed Canary 170
Kashmir 168
Red-throated 168
Rusty-tailed 168
Slaty-blue 170
Spotted 168
Ultramarine 170
Verditer 170
Forktail Little 176
Spotted 176
Francolin Black 50
Grey 50

Gadwall 56
Garganey 58
Godwit Bar-tailed 96
Black-tailed 96
Goldcrest 200
Goldeneye Common 60
Goldfinch Eurasian 226

Goose Bar-headed 54
Greater White-fronted 54
Greylag 54
Lesser White-fronted* 234
Goshawk Northern 128
Grassbird Bristled* 239
Striated 194
Grebe Black-necked 140
Great Crested 140
Horned* 237
Little 140
Red-necked* 237
Greenfinch Yellow-breasted 226
Greenshank Common 96
Griffon Eurasian 122
Himalayan 122
Grosbeak Black-and-yellow 230
Spot-winged 230
White-winged 230
Gull Black-headed 112
Brown-headed 112
Caspian 110
Heuglin's 110
Mew 112
Pallas's 110
Slender-billed 112
Sooty 112

Harrier Eurasian Marsh 124
Hen 126
Montagu's 126
Pallid 126
Pied* 236
Hawfinch 230
Heron Black-crowned Night 144
Grey 142
Indian Pond 144
Little 144
Purple 142
Hobby Eurasian 138
Honey-buzzard Oriental 130
Honeyguide Yellow-rumped* 235
Hoopoe Common 66
Hornbill Indian Grey 66
Oriental Pied* 235
Hypocolius Grey 190

Ibis Black 146
Black-headed 146
Glossy 146

Ibisbill 102

Jacana Bronze-winged 102
Pheasant-tailed 102
Jackdaw Eurasian 156
Jaeger Parasitic 110
Pomarine 110
Jay Black-headed 154
Eurasian 154

Kestrel Common 136
Kingfisher Black-capped* 235
Common 68
Crested 68
Pied 68
White-throated 68
Kite Black 118
Black-shouldered 118
Brahminy 118
Knot Great 98
Red* 235
Koel Asian 72

Lammergeier 120
Lapwing Northern 106
Red-wattled 106
Sociable 106
White-tailed 106
Yellow-wattled 106
Lark Ashy-crowned Sparrow 208
Bar-tailed 208
Bimaculated 208
Black-crowned Sparrow 208
Crested 210
Desert 208
Greater Hoopoe 208
Greater Short-toed 210
Horned 210
Hume's Short-toed 210
Lesser Short-toed 210
Rufous-tailed 208
Sand 210
Laughingthrush Chestnut-crowned* 239
Rufous-chinned 204
Streaked 204
Variegated 204
White-throated 204
Leiothrix Red-billed 206
Linnet Eurasian 226
Loon Red-throated* 238

Magpie Black-billed 154
Yellow-billed Blue 154

Malkoha Sirkeer 72
Mallard 56
Martin Asian House 190
  Dusky Crag 188
  Eurasian Crag 188
  Northern House 190
  Pale 188
  Plain 188
  Rock 188
Merganser Common 60
  Red-breasted 60
Merlin 136
Minivet Long-tailed 158
  Rosy 158
  Small 158
  White-bellied* 238
Monal Himalayan 52
Monarch Black-naped* 238
Moorhen Common 90
Munia Scaly-breasted 224
Myna Bank 182
  Common 182
  Jungle 182

Needletail White-throated 74
Nightingale Common* 239
Nightjar Eurasian 80
  Grey 80
  Indian 80
  Large-tailed 80
  Savanna 80
  Sykes's 80
Niltava Rufous-bellied 170
Noddy Brown 116
Nutcracker Spotted 154
Nuthatch Chestnut-bellied 184
  Eastern Rock 184
  Kashmir 184
  White-cheeked 184

Openbill Asian* 237
Oriole Eurasian Golden 158
Osprey 118
Owl Barn 78
  Boreal* 235
  Brown Fish 78
  Collared Scops 76
  Dusky Eagle 78
  Eurasian Eagle 78
  Eurasian Scops 76
  Little 76
  Long-eared 78
  Mountain Scops 76
  Oriental Scops 76
  Pallid Scops 76
  Short-eared 78
  Snowy* 235
  Tawny 78
Owlet Asian Barred 76
  Collared 76
  Spotted 76
Oystercatcher Eurasian 102

Painted-snipe Greater 94
Paradise-flycatcher Asian 160
Parakeet Alexandrine 72
  Plum-headed 72
  Rose-ringed 72
  Slaty-headed 72
Parrotbill Bearded* 239
Partridge See-see 48
  Snow 48
Peafowl Indian 52
Pelican Dalmatian 150
  Great White 150
Petronia Chestnut-shouldered 214
Phalarope Red* 236
  Red-necked 100
Pheasant Cheer 52
  Kalij 52
  Koklass 50
Piculet Speckled 62
Pigeon Common Wood 82
  Hill 82
  Orange-breasted Green* 235
  Rock 82
  Snow 82
  Speckled Wood 82
  Wedge-tailed Green 84
  Yellow-eyed 82
  Yellow-footed Green 84
Pintail Northern 58
Pipit Buff-bellied 220
  Long-billed 218
  Meadow 220
  Olive-backed* 240
  Paddyfield 218
  Red-throated 220
  Richard's* 240
  Rosy 220
  Tawny 218
  Tree 220
  Upland 218
  Water 220
Pitta Indian 152

Plover Common Ringed 104
  European Golden* 236
  Greater Sand 104
  Grey 104
  Kentish 104
  Lesser Sand 104
  Little Ringed 104
  Pacific Golden 104
Pochard Baer's* 234
  Common 60
  Ferruginous 60
  Red-crested 60
Pratincole Collared 108
  Oriental* 236
  Small 108
Prinia Ashy 192
  Graceful 194
  Grey-breasted 192
  Plain 192
  Rufous-fronted 192
  Rufous-vented 192
  Striated 192
  Yellow-bellied 192
Pygmy-goose Cotton 58

Quail Common 48
  Rain 48

Rail Water 90
Raven Brown-necked 156
  Common 156
Redshank Common 96
  Spotted 96
Redstart Black 174
  Blue-capped 174
  Blue-fronted 176
  Common 174
  Plumbeous Water 176
  Rufous-backed 174
  White-bellied 176
  White-capped Water 176
  White-winged 174
Redwing* 238
Robin Eurasian* 238
  Golden Bush 172
  Indian 174
  Indian Blue 172
  Orange-flanked Bush 172
  Oriental Magpie 174
  Rufous-tailed Scrub 172
Rock-chat Brown 178
Roller European 66
  Indian 66
Rook 156
Rosefinch Common 228
  Great 228

Pink-browed 228
Red-fronted 228
Red-mantled 228
White-browed 228
Rubythroat White-tailed 172
Ruff 100

Sanderling 98
Sandgrouse Black-bellied 92
Chestnut-bellied 92
Crowned 92
Lichtenstein's 92
Painted 92
Pin-tailed 92
Spotted 92
Sandpiper Broad-billed 100
Common 98
Curlew 100
Green 98
Marsh 96
Sharp-tailed* 236
Terek 98
Wood 98
Scaup Greater* 234
Serin Fire-fronted 226
Shearwater Persian 150
Short-tailed* 238
Shelduck Common 56
Ruddy 56
Shikra 128
Shoveler Northern 58
Shrike Bay-backed 152
Great Grey* 238
Long-tailed 152
Red-backed 152
Rufous-tailed 152
Southern Grey 152
Woodchat* 238
Sibia Rufous 206
Silverbill Indian 224
Skimmer Indian 108
Skylark Eurasian 210
Oriental 210
Smew 60
Snipe Common 94
Jack 94
Pintail 94
Solitary 94
Snowcock Himalayan 50
Sparrow Dead Sea 214
Eurasian Tree 214
House 212
Rock 214
Russet 214
Sind 212

Spanish 212
Sparrowhawk Eurasian 128
Spoonbill Eurasian 146
Starling Asian Pied 182
Brahminy 182
Chestnut-tailed 182
Common 182
Purple-backed* 239
Rosy 182
Stilt Black-winged 102
Stint Little 98
Temminck's 98
Stonechat Common 178
White-tailed 178
Stork Black 148
Black-necked 148
Painted 148
White 148
Woolly-necked 148,* 237
Storm-petrel Wilson's 150
Sunbird Crimson 212
Mrs Gould's 212
Purple 212
Swallow Barn 188
Red-rumped 188
Streak-throated 190
Wire-tailed 188
Swamphen Purple 88
Swan Mute* 234
Tundra* 234
Whooper* 234
Swift Alpine 74
Common 74
Fork-tailed 74
House 74
Pallid 74

Tailorbird Common 194
Teal Baikal* 234
Common 58
Tern Black-bellied 116
Bridled 116
Caspian 114
Common 114
Great Crested 114
Gull-billed 112
Lesser Crested 114
Little 116
River 114
Sandwich 114
Saunders's 116
Whiskered 116
White-cheeked 114
White-winged 116
Thick-knee Eurasian 102
Great 102

Thrush Blue Rock 162
Blue Whistling 162
Blue-capped Rock 162
Chestnut 166
Chestnut-bellied Rock 162
Dark-throated 166
Dusky 166
Mistle 166
Orange-headed 162
Plain-backed 164
Rufous-tailed Rock 162
Scaly 164
Song* 238
Tickell's 164
Tit Azure* 239
Black-lored 186
Black-throated 186
Fire-capped 186
Great 186
Green-backed 186
Rufous-naped 186
Spot-winged 186
White-cheeked 186
White-crowned Penduline 186
White-throated 186
Yellow-breasted* 239
Tragopan Western 50
Treecreeper Bar-tailed 184
Eurasian 184
Treepie Grey 154
Rufous 154
Tropicbird Red-billed 150
Turnstone Ruddy 100
Twite 226

Vulture Cinereous 122
Egyptian 120
Indian 120
Red-headed* 236
White-rumped 120

Wagtail Citrine 216
Forest 216
Grey 218
White 216
White-browed 216
Yellow 216
Wallcreeper 184
Warbler Barred* 239
Blunt-winged 198
Blyth's Leaf* 239
Blyth's Reed 198
Booted 200
Brooks's Leaf 202

Brownish-flanked Bush 198
Cetti's Bush 198
Clamorous Reed 198
Desert 196
Dusky* 239
Eurasian Reed* 239
Grasshopper 198
Great Reed* 239
Greenish 202
Grey-hooded 202
Grey-sided Bush 198
Hume's 202
Large-billed Leaf 202
Lemon-rumped 202
Long-billed Bush 198
Ménétries 196
Moustached 198
Orphean 196
Paddyfield 198
Plain Leaf 200
Streaked Scrub 194
Sulphur-bellied 200
Tickell's Leaf 200
Tytler's Leaf 202
Upcher's 200
Western Crowned 202
Whistler's 202
White-browed Tit 194
Watercock 88
Waterhen White-breasted 90
Waxwing Bohemian* 238
Weaver Baya 224
  Black-breasted 224
  Streaked 224
Wheatear Desert 180
  Finsch's 180
  Hooded 178
  Hume's 178
  Isabelline 180
  Northern 180
  Pied 180
  Rufous-tailed 180
  Variable 180
Whimbrel 96
Whistling-duck Fulvous* 234
  Lesser 54
White-eye Oriental 194
Whitethroat Greater 196
  Lesser 196
Wigeon Eurasian 56
Woodcock Eurasian 94
Woodpecker Brown-fronted 62
  Fulvous-breasted 62
  Grey-capped Pygmy 62
  Grey-headed 64
  Himalayan 64
  Rufous-bellied 62
  Scaly-bellied 64
  Sind 64
  Yellow-crowned 62
Woodshrike Common 160
Wren Winter 184
Wryneck Eurasian 62

Yellownape Lesser* 235

Scientific Names

Accipiter badius 128
  gentilis 128
  nisus 128
  *virgatus*** 236
Acridotheres fuscus 182
  ginginianus 182
  tristis 182
Acrocephalus agricola 198
  *arundinaceus*** 239
  concinens 198
  dumetorum 198
  melanopogon 198
  *scirpaceus** 239
  stentoreus 198
Actitis hypoleucos 98
Aegithalos concinnus 186
  leucogenys 186
  niveogularis 186
*Aegolius funereus*** 235
Aegypius monachus 122
Aethopyga gouldiae 212
  siparaja 212
Alaemon alaudipes 208
Alauda arvensis 210
  gulgula 210
Alcedo atthis 68
Alectoris chukar 48
Amandava amandava 224
Amaurornis akool 90
  phoenicurus 90
Ammomanes cincturus 208
  deserti 208
  phoenicurus 208
Ammoperdix griseogularis 48
Anas acuta 58
  clypeata 58
  crecca 58
  *falcata*** 234
  *formosa** 234
  penelope 56
  platyrhynchos 56
  poecilorhyncha 56
  querquedula 58
  strepera 56
*Anastomus oscitans** 237
Anhinga melanogaster 140
Anous stolidus 116
Anser albifrons 54
  anser 54
  *erythropus** 234
  indicus 54
*Anthracoceros albirostris** 235
Anthus campestris 218
  cervinus 220
  *hodgsoni** 240
  pratensis 220
  *richardi** 240
  roseatus 220
  rubescens 220
  rufulus 218
similis 218
spinoletta 220
sylvanus 218
trivialis 220
Apus affinis 74
  apus 74
  pacificus 74
  pallidus 74
Aquila chrysaetos 132
  clanga 132
  *hastata** 237
  heliaca 134
  nipalensis 132
  rapax 132
Ardea cinerea 142
  purpurea 142
Ardeola grayii 144
Ardeotis nigriceps 86
Arenaria interpres 100
Asio flammeus 78
  otus 78
Athene brama 76
  noctua 76
*Aythya baeri** 234
  ferina 60
  fuligula 60
  *marila** 234
  nyroca 60

*Bombycilla garrulus*** 238

Botaurus stellaris 144
Bradypterus major 198
Bubo bubo 78
   coromandus 78
Bubulcus ibis 142
Bucanetes githagineus 228
   mongolicus 228
Bucephala clangula 60
Burhinus oedicnemus 102
Butastur teesa 130
Buteo buteo 130
   rufinus 130
Butorides striatus 144

Cacomantis passerinus 72
Calandrella acutirostris 210
   brachydactyla 210
   raytal 210
   rufescens 210
Calidris acuminata* 236
   alba 98
   alpina 100
   canutus* 235
   ferruginea 100
   minuta 98
   temminckii 98
   tenuirostris 98
Callacanthis burtoni 226
Caprimulgus affinis 80
   asiaticus 80
   europaeus 80
   indicus 80
   macrurus 80
   mahrattensis 80
Carduelis cannabina 226
   carduelis 226
   flavirostris 226
   spinoides 226
Carpodacus erythrinus 228
   puniceus 228
   rhodochlamys 228
   rodochrous 228
   rubicilla 228
   thura 228
Casmerodius albus 142
Catreus wallichii 52
Centropus sinensis 72
Cephalopyrus flammiceps 186
Cercomela fusca 178
Cercotrichas galactotes 172
Certhia familiaris 184
   himalayana 184
Ceryle rudis 68
Cettia brunnifrons 198
   cetti 198
   fortipes 198
Chaetornis striatus* 239

Chaimarrornis leucocephalus 176
Charadrius alexandrinus 104
   dubius 104
   hiaticula 104
   leschenaultii 104
   mongolus 104
   morinellus* 236
Chlamydotis macqueenii 86
Chlidonias hybridus 116
   leucopterus 116
Chrysomma altirostre 204
   sinense 204
Ciconia ciconia 148
   episcopus 148,* 237
   nigra 148
Cinclus cinclus 160
   pallasii 160
Circaetus gallicus 124
Circus aeruginosus 124
   cyaneus 126
   macrourus 126
   melanoleucos* 236
   pygargus 126
Cisticola juncidis 194
Clamator jacobinus 70
Clangula hyemalis* 234
Coccothraustes coccothraustes 230
Columba eversmanni 82
   hodgsonii 82
   leuconota 82
   livia 82
   palumbus 82
   rupestris 82
Copsychus saularis 174
Coracias benghalensis 66
   garrulus 66
Coracina macei 158
   melaschistos 158
Corvus corax 156
   corone 156
   frugilegus 156
   macrorhynchos 156
   monedula 156
   ruficollis 156
   splendens 156
Coturnix coromandelica 48
   coturnix 48
Crex crex* 235
Cuculus canorus 70
   micropterus 70
   poliocephalus 70
   saturatus 70
Culicicapa ceylonensis 170
Cursorius coromandelicus 108
   cursor 108

Cygnus columbianus* 234
   cygnus* 234
   olor* 234
Cyornis rubeculoides 170

Delichon dasypus 190
   urbica 190
Dendrocitta formosae 154
   vagabunda 154
Dendrocopos assimilis 64
   auriceps 62
   canicapillus 62
   himalayensis 64
   hyperythrus 62
   macei 62
   mahrattensis 62
Dendrocygna bicolor* 234
   javanica 54
Dendronanthus indicus 216
Dicaeum agile 212
Dicrurus leucophaeus 160
   macrocercus 160
Dinopium benghalense 64
Dromas ardeola 102
Dupetor flavicollis 144

Egretta garzetta 142
   gularis 142
Elanus caeruleus 118
Emberiza aureola* 240
   bruniceps 232
   buchanani 232
   cia 230
   fucata 232
   hortulana* 240
   leucocephalos 230
   melanocephala 232
   rutila* 240
   schoeniclus 232
   stewarti 232
   striolata 232
Enicurus maculatus 176
   scouleri 176
Ephippiorhynchus asiaticus 148
Eremophila alpestris 210
Eremopterix grisea 208
Eremopterix nigriceps 208
Erithacus rubecula* 238
Esacus recurvirostris 102
Eudynamys scolopacea 72
Eumyias thalassina 170

Falco cherrug 138
   chicquera 136
   columbarius 136
   concolor 136
   jugger 138

peregrinus 138
  subbuteo 138
  tinnunculus 136
Ficedula parva 168
  subrubra 168
  superciliaris 170
  tricolor 170
Francolinus francolinus 50
  pondicerianus 50
Fringilla coelebs 224
  montifringilla 224
Fulica atra 90

Galerida cristata 210
Gallicrex cinerea 88
Gallinago gallinago 94
  solitaria 94
  stenura 94
Gallinula chloropus 90
Garrulax albogularis 204
  erythrocephalus* 239
  lineatus 204
  rufogularis 204
  variegatus 204
Garrulus glandarius 154
  lanceolatus 154
Gavia stellata* 238
Gelochelidon nilotica 112
Glareola lactea 108
  maldivarum* 236
  pratincola 108
Glaucidium brodiei 76
  cuculoides 76
Grus antigone 88
  grus 88
  virgo 88
Gypaetus barbatus 120
Gyps bengalensis 120
  fulvus 122
  himalayensis 122
  indicus 120

Haematopus ostralegus 102
Halcyon pileata* 235
  smyrnensis 68
Haliaeetus albicilla 118
  leucoryphus 118
Haliastur indus 118
Heterophasia capistrata 206
Hieraaetus fasciatus 134
  pennatus 134
Hierococcyx sparverioides* 235
  varius 70
Himantopus himantopus 102
Hippolais caligata 200
  languida 200
Hirundapus caudacutus 74

Hirundo concolor 188
  daurica 188
  fluvicola 190
  fuligula 188
  rupestris 188
  rustica 188
  smithii 188
Hodgsonius phaenicuroides 176
Hydrophasianus chirurgus 102
Hypocolius ampelinus 190
Hypothymis azurea* 238
Hypsipetes leucocephalus 190

Ibidorhyncha struthersii 102
Ictinaetus malayensis 124
Indicator xanthonotus* 235
Ixobrychus cinnamomeus 144
  minutus 144
  sinensis 144

Jynx torquilla 62

Ketupa zeylonensis 78

Lanius collurio 152
  excubitor* 238
  isabellinus 152
  meridionalis 152
  schach 152
  senator* 238
  vittatus 152
Larus brunnicephalus 112
  cachinnans 110
  canus 112
  genei 112
  hemprichii 112
  heuglini 110
  ichthyaetus 110
  ridibundus 112
Leiothrix lutea 206
Leptopoecile sophiae 194
Leptoptilos dubius* 237
Lerwa lerwa 48
Leucosticte brandti 222
  nemoricola 222
Limicola falcinellus 100
Limosa lapponica 96
  limosa 96
Locustella naevia 198
Lonchura malabarica 224
  punctulata 224
Lophophorus impejanus 52
Lophura leucomelanos 52
Loxia curvirostra 230
Luscinia brunnea 172
  megarhynchos* 239
  pectoralis 172

svecica 172
Lymnocryptes minimus 94

Marmaronetta angustirostris 58
Megaceryle lugubris 68
Megalaima asiatica 66
  haemacephala 66
  virens 66
Megalurus palustris 194
Melanocorypha bimaculata 208
Melophus lathami 230
Mergellus albellus 60
Mergus merganser 60
  serrator 60
Merops apiaster 68
  orientalis 68
  persicus 68
  philippinus 68
Mesophoyx intermedia 142
Metopidius indicus 102
Miliaria calandra* 240
Milvus migrans 118
Mirafra cantillans 208
  erythroptera 208
Monticola cinclorhynchus 162
  rufiventris 162
  saxatilis 162
  solitarius 162
Motacilla alba 216
  cinerea 218
  citreola 216
  flava 216
  maderaspatensis 216
Muscicapa dauurica 168
  ruficauda 168
  sibirica 168
  striata 168
Mycerobas carnipes 230
  icterioides 230
  melanozanthos 230
Mycteria leucocephala 148
Myophonus caeruleus 162

Nectarinia asiatica 212
Neophron percnopterus 120
Netta rufina 60
Nettapus coromandelianus 58
Niltava sundara 170
Nucifraga caryocatactes 154
Numenius arquata 96
  phaeopus 96
Nyctea scandiaca* 235
Nycticorax nycticorax 144

Oceanites oceanicus 150
Ocyceros birostris 66
Oenanthe alboniger 178

deserti 180
finschii 180
isabellina 180
monacha 178
oenanthe 180
picata 180
pleschanka 180
xanthoprymna 180
Oriolus oriolus 158
Orthotomus sutorius 194
Otis tarda* 235
Otus bakkamoena 76
  brucei 76
  scops 76
  spilocephalus 76
  sunia 76
Oxyura leucocephala 54

Pandion haliaetus 118
Panurus biarmicus* 239
Parus cyanus* 239
  flavipectus 239
  major 186
  melanolophus 186
  monticolus 186
  rufonuchalis 186
  xanthogenys 186
Passer domesticus 212
  hispaniolensis 212
  moabiticus 214
  montanus 214
  pyrrhonotus 212
  rutilans 214
Pavo cristatus 52
Pelecanus crispus 150
  onocrotalus 150
Pericrocotus cinnamomeus 158
  *erythropygius** 238
  ethologus 158
  roseus 158
Pernis ptilorhyncus 130
Petronia petronia 214
  xanthocollis 214
Phaenicophaeus leschenaultii 72
Phaethon aethereus 150
Phalacrocorax carbo 140
  fuscicollis 140
  niger 140
  *pygmeus** 237
Phalaropus fulicaria* 236
  lobatus 100
Philomachus pugnax 100
Phoenicopterus minor 146
  ruber 146
Phoenicurus coeruleocephalus 174
  erythrogaster 174
  erythronota 174
  frontalis 176
  ochruros 174
  phoenicurus 174
Phylloscopus affinis 200
  chloronotus 202
  collybita 200
  *fuscatus** 239
  griseolus 200
  humei 202
  magnirostris 202
  neglectus 200
  occipitalis 202
  *reguloides** 239
  sindianus 200
  subviridis 202
  trochiloides 202
  tytleri 202
Pica pica 154
Picumnus innominatus 62
Picus canus 64
  *chlorolophus** 235
  squamatus 64
Pitta brachyura 152
Platalea leucorodia 146
Plegadis falcinellus 146
Ploceus benghalensis 224
  manyar 224
  philippinus 224
Pluvialis apricaria* 236
  fulva 104
  squatarola 104
Podiceps auritus* 237
  cristatus 140
  *grisegena** 237
  nigricollis 140
Pomatorhinus erythrogenys 204
Porphyrio porphyrio 88
Porzana fusca 90
  parva 90
  porzana 90
  pusilla 90
Prinia buchanani 192
  burnesii 192
  criniger 192
  flaviventris 192
  gracilis 194
  hodgsonii 192
  inornata 192
  socialis 192
Prunella atrogularis 222
  collaris 222
  fulvescens 222
  himalayana 222
  *ocularis** 240
  rubeculoides 222

strophiata 222
Psarisomus dalhousiae* 238
Pseudibis papillosa 146
Psittacula cyanocephala 72
  eupatria 72
  himalayana 72
  krameri 72
Pterocles alchata 92
  coronatus 92
  exustus 92
  indicus 92
  lichtensteinii 92
  orientalis 92
  senegallus 92
Pteruthius flaviscapis 206
  xanthochlorus 206
Pucrasia macrolopha 50
Puffinus persicus 150
  *tenuirostris** 238
Pycnonotus cafer 190
  leucogenys 190
  leucotis 190
Pyrrhocorax graculus 156
  pyrrhocorax 156
Pyrrhula aurantiaca 230

Rallus aquaticus 90
Recurvirostra avosetta 102
Regulus regulus 200
Remiz coronatus 186
Rhipidura albicollis 158
  aureola 158
  hypoxantha 158
Rhodopechys sanguinea 226
Rhodospiza obsoleta 226
Rhyacornis fuliginosus 176
Riparia diluta 188
  paludicola 188
Rostratula benghalensis 94
Rynchops albicollis 108

Sarcogyps calvus* 236
Sarkidiornis melanotos 54
Saxicola caprata 178
  ferrea 178
  leucura 178
  macrorhyncha 178
  torquata 178
Saxicoloides fulicata 174
Scolopax rusticola 94
Scotocerca inquieta 194
Seicercus whistleri 202
  xanthoschistos 202
Serinus pusillus 226
Sitta cashmirensis 184
  castanea 184
  leucopsis 184

tephronota 184
Spilornis cheela* 236
Spizaetus nipalensis 134
Stachyris pyrrhops 204
Stercorarius parasiticus 110
  pomarinus 110
Sterna acuticauda 116
  albifrons 116
  anaethetus 116
  aurantia 114
  bengalensis 114
  bergii 114
  caspia 114
  hirundo 114
  repressa 114
  sandvicensis 114
  saundersi 116
Streptopelia chinensis 84
  decaocto 84
  orientalis 84
  senegalensis 84
  tranquebarica 84
  turtur 84
Strix aluco 78
Sturnus contra 182
  malabaricus 182
  pagodarum 182
  roseus 182
  sturninus* 239
  vulgaris 182
Sula dactylatra 150
Sylvia communis 196

  curruca 196
  hortensis 196
  mystacea 196
  nana 196
  nisoria* 239
Sypheotides indica 86

Tachybaptus ruficollis 140
Tachymarptis melba 74
Tadorna ferruginea 56
  tadorna 56
Tarsiger chrysaeus 172
  cyanurus 172
Tephrodornis pondicerianus 160
Terpsiphone paradisi 160
Tetraogallus himalayensis 50
Tetrax tetrax 86
Threskiornis melanocephalus 146
Tichodroma muraria 184
Tragopan melanocephalus 50
Treron bicincta* 235
  phoenicoptera 84
  sphenura 84
Tringa erythropus 96
  glareola 98
  nebularia 96
  ochropus 98
  stagnatilis 96
  totanus 96
Troglodytes troglodytes 184

Turdoides caudatus 206
  earlei 206
  malcolmi 206
  striatus 206
Turdus boulboul 164
  iliacus* 238
  merula 164
  naumanni 166
  philomelos* 238
  rubrocanus 166
  ruficollis 166
  unicolor 164
  viscivorus 166
Turnix sylvatica 48
  tanki 48
Tyto alba 78

Upupa epops 66
Urocissa flavirostris 154

Vanellus gregarius 106
  indicus 106
  leucurus 106
  malarbaricus 106
  vanellus 106

Xenus cinereus 98

Zoothera citrina 162
  dauma 164
  mollissima 164
Zosterops palpebrosus 194